Generative AI for Business Innovation

A Practical Guide for Leadership to Realize Business Value with Generative AI

Brajesh De

Apress®

Generative AI for Business Innovation: A Practical Guide for Leadership to Realize Business Value with Generative AI

Brajesh De
Bangalore, Karnataka, India

ISBN-13 (pbk): 979-8-8688-2681-8 ISBN-13 (electronic): 979-8-8688-2682-5
https://doi.org/10.1007/979-8-8688-2682-5

Managing Director, Apress Media LLC: Welmoed Spahr
Acquisitions Editor: Celestin Suresh John
Development Editor: Laura Berendson
Editorial Assistant: Gryffin Winkler

Cover designed by eStudioCalamar

Distributed to the book trade worldwide by Springer Science+Business Media New York, 1 New York Plaza, New York, NY 10004. Phone 1-800-SPRINGER, fax (201) 348-4505, e-mail orders-ny@springer-sbm.com, or visit www.springeronline.com. Apress Media, LLC is a Delaware LLC and the sole member (owner) is Springer Science + Business Media Finance Inc (SSBM Finance Inc). SSBM Finance Inc is a **Delaware** corporation.

For information on translations, please e-mail booktranslations@springernature.com; for reprint, paperback, or audio rights, please e-mail bookpermissions@springernature.com.

Apress titles may be purchased in bulk for academic, corporate, or promotional use. eBook versions and licenses are also available for most titles. For more information, reference our Print and eBook Bulk Sales web page at http://www.apress.com/bulk-sales.

Any source code or other supplementary material referenced by the author in this book is available to readers on GitHub. For more detailed information, please visit https://www.apress.com/gp/services/source-code.

If disposing of this product, please recycle the paper

In loving memory of my parents, whose blessings and spirit remain my guiding light.

Table of Contents

Chapter 5: Building a Winning Generative AI Strategy 155

About the Author

Brajesh De is a tech visionary and CTO-level executive with over 25 years of global experience shaping how enterprises design, scale, and govern digital platforms in an AI-driven world.

Over the course of his career, Brajesh has successfully led several large-scale digital transformation programs for Fortune 500 organizations, spanning strategy, architecture, and execution. He has witnessed firsthand both the promise of Generative AI and the practical challenges of scaling it responsibly and reliably across data, platforms, security, and organizational readiness. As a trusted advisor to boards, CIOs/CTOs, he influences strategic technology investments, modernization roadmaps, and long-term AI-enabled digital transformation agendas—helping organizations transform their dream experiments to reality and derive measurable value using Generative AI.

A recognized industry thought leader, Brajesh is a published author, conference speaker, and builder of Global Capability Centers (GCCs) and Center of Excellence (CoEs). He holds patents in API assessment and data veracity across four countries, reflecting his long-standing focus on building trustworthy and scalable digital systems.

Brajesh received his bachelor's degree in Electrical Engineering from IIT-BHU, Varanasi, and was the recipient of multiple university gold medals during his tenure.

About the Technical Reviewer

 **Laxmi Vanam** is a data and analytics professional with 16+ years of experience driving AI-powered modernization, data strategy, and self-service BI initiatives across finance, insurance, and technology sectors. She specializes in bridging business needs with technical execution, leveraging expertise in data architecture, cloud analytics, and applied machine learning. As an Apress reviewer, she brings a deep appreciation for clarity, precision, and real-world relevance in technical writing.

Acknowledgments

I am deeply grateful to everyone who helped and supported me in bringing this book to life. Your encouragement, patience, and belief made all the difference.

Thank you.

Structure of the Book

The book guides you through the journey with **three progressive parts.**

Part 1: The Business Leader's Guide to Generative AI

Lays the foundation. It demystifies Generative AI, providing leaders with conceptual clarity and strategic context.

- **Chapter 1: Demystifying Generative AI**

  Explains how Generative AI works, covering its benefits and industry applications. It deconstructs the technology powering Generative AI using simple language designed for decision-makers and not just engineers.

- **Chapter 2: A Double-Edged Sword: Benefits and Risks of Generative AI**

  Explores both the extraordinary opportunities and the real risks-bias, hallucinations, security, compliance, and ethical challenges, covering ways to address them. The chapter emphasizes why responsible AI is a non-negotiable requirement for AI success.

- **Chapter 3: Generative AI in Action: Industry Use Cases**

  Explores how industries are using Generative AI to create new products, enhance customer experiences, streamline operations, and gain a competitive advantage.

- **Chapter 4: Assessing Your Generative AI Readiness**

  Provides a structured framework for evaluating organizational maturity-across leadership alignment, data quality, talent capability, infrastructure, governance, and culture.

At the end of this section, you will have a good understanding of Generative AI technologies, the business use cases that can be built, and your organization's readiness in the adoption journey.

Part 2: Building Your Generative AI Strategy

This section shifts from theory to operationalization. This section covers in detail the strategy and pillars for successfully taking a Generative AI pilot to production.

- **Chapter 5: Building a Winning Generative AI Strategy**

  Covers all the strategic aspects that must be considered at the outset to ensure the successful implementation and rollout of generative AI within the organization.

- **Chapter 6: Generative AI Architecture and Infrastructure: Building the Foundation**

  Covers scalable, secure, and cost-effective architecture patterns, integration approaches, and infrastructure design principles to build responsible and reliable Generative AI solutions.

- **Chapter 7: Building the Data Architecture for Generative AI**

  Highlights the importance of using high-quality, governed, and properly organized data, supported by a strong and modern data architecture, to ensure reliable and trustworthy AI results.

- **Chapter 8: Building a High-Performance Team for Generative AI**

  Discusses talent models, skill sets, operating models, and how to build an AI-first culture that drives enterprise adoption.

This section equips leaders with the strategic blueprint and technical foundations necessary to move beyond isolated experiments toward enterprise capability.

Part 3: Generative AI: Shaping the Future

Focuses on scale and sustainability and how to be prepared to adopt the next evolution of AI.

- **Chapter 9: Scaling Generative AI: From Pilot to Enterprise Value**

  Outlines best practices for transitioning from proofs of concept and pilots to secure, production-grade deployments with measurable ROI.

- **Chapter 10: The Next Wave of Generative AI: Preparing Business and Technology Leadership for the Future**

  Explores emerging trends, evolving architectures, governance evolution, and how leaders must prepare for what comes next.

This section prepares leaders not only to adopt Generative AI but also to evolve with it.

What You Will Gain From This Book

By the time you finish this book, you will not only have a deep understanding of Generative AI concepts and best practices but also gain practical clarity, strategic guidance, and confidence to execute.

Introduction

Generative AI: A Defining and Game-Changing Moment for Business Leaders

Technology has been evolving at a rapidly increasing pace. In the last few decades, with the evolution of the internet, cloud computing and mobile technology, businesses have fundamentally changed how they compete, innovate, and create value. Internet refined connectivity. Cloud computing redefines scalability and resilience. Mobile captivated users through high-tech digital experiences.

*Now, **Generative AI** is rewriting the rules of intelligence.*

Generative AI enhances human creativity, accelerates decision-making, and enables entirely new business models. It can write, design, code, analyze, reason, and interact in ways that were earlier considered as "human-only" capabilities. It is not another utility or a tool in the technology stack. It is bringing in a structural shift by laying a new foundational capability that can redefine and reshape products, operational workflows and competitive strategy.

Amid the landscape of hype, confusion, and fragmented pilots, executives and business leaders are looking for a clear path for Generative AI. Leadership is seeking an integrated strategy to steer them through this evolving landscape. They want guidance on how to adopt Generative AI for their business in production and realize a measurable return on their investments. Executives are asking

- Where should they start their Generative AI journey?

- What are the real business use cases in their industry that offer real Gen AI value?

- How do we mitigate risks and govern responsibly?

- How do we scale from proof-of-concept to enterprise value?

- How do we align Generative AI with long-term strategy?

This book is written to answer those questions.

Who Is This Book For

Generative AI for Business Innovation is designed for

- **Business leaders and CXOs** who must make strategic investment decisions

- **Technology leaders and architects** responsible for translating vision into scalable systems

- **Innovation heads and transformation leaders** driving digital reinvention

- **Entrepreneurs and product leaders** seeking new business models powered by AI

While exploring the foundational concepts and architectural considerations for Generative AI, this book focuses on **strategic clarity, practical execution, and sustainable value creation.** It bridges the gap between strategic intent and technical feasibility.

What This Book Is About

At its heart, this book emphasizes one central theme:

Generative AI is not a technology initiative. It is the foundational capability for all business transformation initiatives.

This book provides a practical end-to-end roadmap-from understanding the fundamentals to building enterprise-grade Generative AI capabilities that are secure, scalable, ethical, and value-driven. It approaches Generative AI from a business-first perspective and addresses the following questions:

- How can you use Generative AI for business advantage?

- How can organizations assess their true readiness for Generative AI?

- What should be the constituents of a winning Generative AI strategy?

- What architecture and data foundations are required to scale?

- How to transition from pilot to production while minimizing risks?

- How to address safety and reliability concerns with a Responsible AI philosophy?

- How should leaders prepare for the next wave of AI innovation?

CHAPTER 1

Demystifying Generative AI

Introduction

In today's digitally unified world, business is going through a seismic shift in the way it operates, innovates, and competes. Generative AI is fundamentally redefining the way we create, innovate, and operate and is at the heart of all innovative business transformations. According to a report published by McKinsey Global Institute, **by 2030, 70% of companies that fail to integrate Generative AI will cease to be competitive in their markets**. Hence, companies across all industries must invest in adopting Generative AI.

Traditional AI has been mostly used to detect patterns and make predictions. Generative AI takes a leap further with its revolutionary capabilities to interact with human beings in natural language and generate new outputs that are customized for the users. This opens up lots of opportunities for businesses across industries.

From hyper-personalized customer experiences and AI-driven product design to automated content generation and real-time business intelligence, Generative AI is transforming how businesses operate. According to a report from Accenture, 97% of organizations expect Generative AI to be transformative. It can streamline processes, automate

B. De, *Generative AI for Business Innovation*, https://doi.org/10.1007/979-8-8688-2682-5_1

complex tasks, spark creativity, boost productivity, and accelerate growth. With an estimated annual impact of \$2.6 to \$4.4 trillion across industries—and a projected \$15.7 trillion contribution to the global economy by 2030, according to PwC and BCG—its business potential is immense.

Generative AI is no longer a futuristic concept. It is a present day reality enabling organizations to scale innovation at an unprecedented speed, disrupt markets, and unlock new opportunities. Companies that effectively harness Generative AI will have a significant competitive advantage.

Key Takeaway: *Generative AI is reshaping the future of business by not only automating tasks but enabling creativity, innovation, and scale. Businesses that fail to adopt Generative AI may lose the competitive edge.*

Basics of Artificial Intelligence (AI)

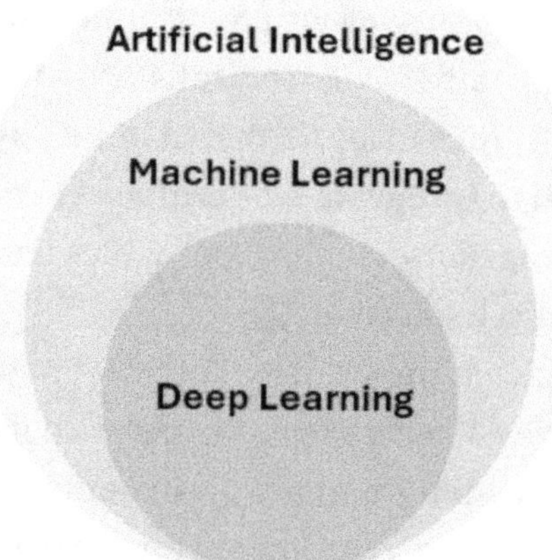

Figure 1-1. *Constituents of Artificial Intelligence*

Artificial Intelligence

Artificial Intelligence, abbreviated as AI, is a discipline or a branch of computer science that deals with the creation of intelligent systems that can learn, reason, and act autonomously like human beings. These systems, also called AI agents, can perform tasks that require human intelligence. Figure 1-1 illustrates the relationship between Artificial Intelligence, Machine Learning and Deep Learning.

Machine Learning

Machine Learning (ML) is a subfield of AI that focuses on using data and algorithms to train a computer program that can imitate the way that humans learn, think, and respond. This trained software, often referred to as the **AI model**, can make predictions for new data without explicit programming based on the patterns it learned. You can think of the AI model as a student learning from textbooks (data). An AI model is made of training code and data. The mathematical representation of an AI model is shown in Figure 1-2.

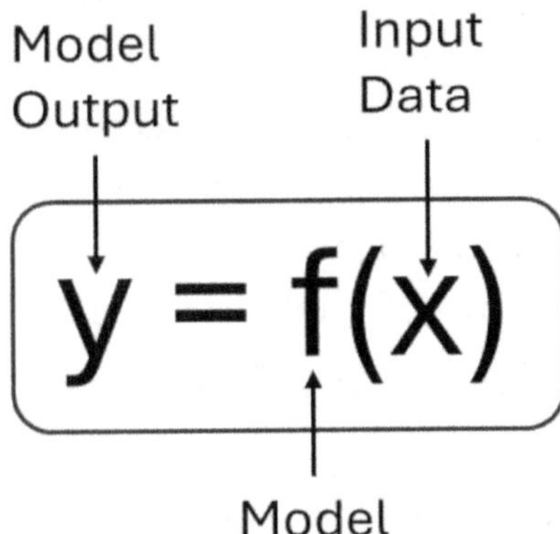

Figure 1-2. *Mathematical Representation of Model*

Machine learning models can be of different types based on the data used for training. If the training data is labeled by humans, then the model is called a **supervised model.** If the model is trained on data without human supervision, then it is called an **unsupervised model**. Figure 1-3 summarizes the major categories of machine learning.

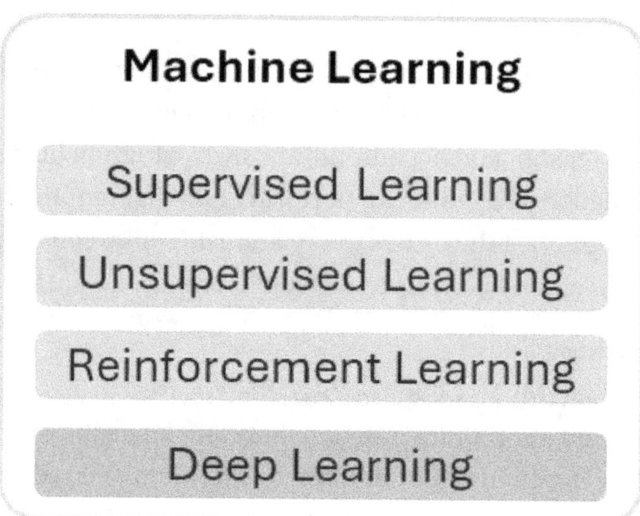

Figure 1-3. *Types of Machine Learning*

Supervised Learning

Supervised learning models learn from past data that is labeled. Data is labeled with a tag like a name, type, or number. Supervised learning models can be used to predict, classify, and cluster data based on the patterns it has learnt from the labeled data. For example, if a supervised model is trained based on the data for city, date, and past climatic conditions like temperature and rainfall, then this model can be used to predict the probable climatic conditions for a date in the future.

Unsupervised Learning

Unsupervised learning is a type of machine learning that learns from data without human supervision. Unlike supervised models, unsupervised machine learning models learn on their own from data that doesn't have any labels or tags. It analyzes the data to discover patterns, relationships, and insights without any explicit guidance or instruction. It can find similarities and differences in information, which makes it ideal for exploratory data analysis, cross-selling strategies, customer segmentation, and image recognition.

Reinforcement Learning

Reinforcement learning is a type of machine learning where autonomous agents learn to make decisions and act in response to the environment. The agents are trained using rewards and punishment mechanisms. They take the best possible action that gives maximum reward and minimum punishment in a given environment or situation. Essentially, the autonomous agent learns to perform a task through trial and error without direct instructions from any human user. They can solve problems without predefined solutions or explicitly programmed actions and most importantly, without large amounts of data. This makes it suitable for implementing solutions for self-driving cars, robotics, and gaming.

Deep Learning

Deep Learning is a subset of machine learning that uses artificial neural networks to process more complex patterns. It is inspired by the human brain that is made of many interconnected nodes, or neurons. These nodes or neurons can learn to perform tasks by processing data and making predictions. Deep learning models have many layers of neurons that allow them to learn more complex patterns. Neural networks can use both labeled and unlabeled data for their learning. This is called semi-supervised learning. In semi-supervised learning, the neural network is generally trained on a small amount of labeled data and a large amount of unlabeled data. The labeled data helps to learn the basic concepts of the task, while the unlabeled data helps the neural network to generalize new examples.

Deep learning models are of two types—discriminative model and generative model. **Discriminative model** is used for classification and prediction based on the relationship or patterns learned from the dataset. On the other hand, a **generative model** produces new data or content that is similar to the data it was trained on. It understands the distribution of data and can generate the next word in the sequence. A discriminative model can be used to identify the animal in a picture. For example, whether the picture is that of a dog or a lion. Whereas a Generative Model can generate a new picture of an animal based on the description provided. For example, create a picture of a roaring lion. Hence, to summarize, a discriminative model discriminates between different types of data instances and outputs a type, number, probability, or a class. Whereas, a generative model can be used to create new data instances or content that can be a natural language text, image, audio, or video or even a new design. Today transformer-based models like GPT power most AI applications.

Foundation Model

A Foundation Model is a large AI Model built using deep neural networks and pre-trained on a vast quantity of labeled and unlabeled data. Since it is pre-trained, it can perform a wide range of generalized tasks more quickly and cost effectively. It can perform common tasks like question and answer, translation, text summarization, sentiment analysis, image classification and generation, object recognition, natural language conversation, and more. Foundation models have the potential to revolutionize many industries with their capability to perform a wide range of tasks. They can be used for many complex activities like fraud detection, personalized customer support, robotics, healthcare, autonomous vehicles, and more. BERT, GPT, Claude, DALL-E, Cohere, Llama, BLOOM are some examples of well-known foundation models.

Key Takeaway: *AI enables machines to perform tasks like humans; Machine Learning helps machines to learn from data. Deep Learning enables machine to recognize complex patterns.*

Introducing Generative AI

Generative AI marks a revolutionary shift—from understanding data to creating it. It is a class of artificial intelligence that can generate new content and ideas, like human beings, based on what it has learnt from existing content in response to a query or prompt. The content can be a text, an image, an audio/music, a video, or any other type of data like a product design. Figure 1-4 shows the multi-modal generation capabilities of a Foundation model.

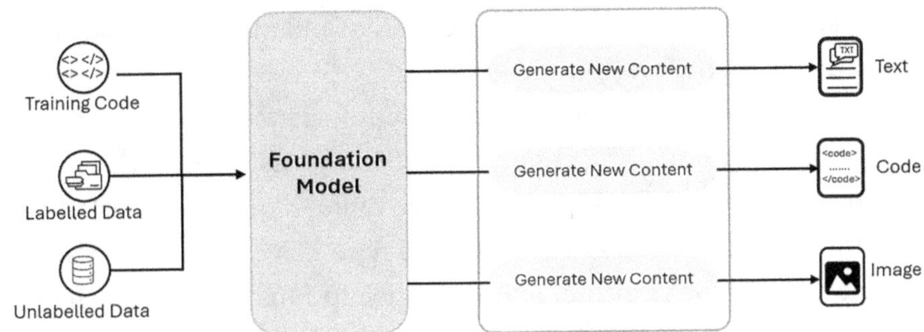

Figure 1-4. *The Foundation Model Paradigm—Enabling Multi-Modal Generation*

Generative AI uses foundation models based on neural networks. The neural network makes Generative AI a highly powerful tool to create new and original contents like humans. The foundation models are trained on large volumes of existing content to learn human language, programming languages, art, chemistry, biology, or any complex subject matter. Generative AI uses the trained foundation model to respond with newly generated content in response to inputs given as a "prompt."

The trained Generative AI model can be used to solve new problems. For example, it can learn English vocabulary and write poetry or an essay, create a blog post, or respond to emails. Based on learnings from image-based training data, it can generate new artwork. Businesses can use Generative AI for various purposes, like draft reports, personalize marketing campaigns, customer service chatbots, product development and design. Software vendors are integrating Generative AI into their products to boost productivity and improve decision-making. Generative AI is also being used to create synthetic data to train other AI and machine learning models.

Traditional AI vs. Generative AI

Traditional AI is a subset of artificial intelligence technology. It is designed to perform tasks based on predetermined algorithms and data. It learns patterns from training data sets and applies that learning to identify similar patterns elsewhere. This makes traditional AI more reactive. Consequently, it is well-suited for executing specific repetitive and mundane tasks that require intelligence derived from learning.

By analyzing data, traditional AI can identify patterns, make decisions, offer recommendations, and generate predictions. Its applications include fraud detection, spam filtering, image recognition, language translation, forecasting, complex problem-solving, and data-driven decision-making.

Generative AI, on the other hand, is an evolution of AI to bring in the creativity of human minds in machines. This is a game changer that can bring in endless possibilities. Generative AI models are trained on a massive amount of data. That learning and analysis is then used to not just recognize patterns and make future predictions but also create new contents. This new content is generated based on human inputs—a command or a set of guidelines or parameters often referred to as "***prompt***." The prompts can be in any form—text, graphics, or audio.

One big difference from traditional AI is that Generative AI uses neural networks to create new and unique content such as text, image, and music. This generative ability makes it even more human and natural to interact and use. Figure 1-5 shows the decision tree to identify when an ML model is generative or non-generative.

Generative AI can be used to create personalized, just-in-time marketing content, email campaigns to drive customer loyalty. Almost every industry is finding new ways to use Generative AI to implement various use cases that never existed before.

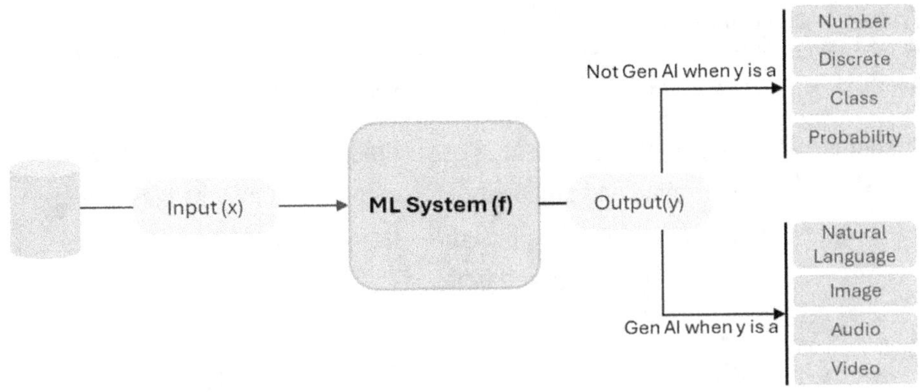

Figure 1-5. *When an AI Model Is Generative AI Model*

Table 1-1 summarizes the key differences between the traditional AI and Generative AI.

Table 1-1. *Differences Between Traditional AI and Generative AI*

Aspects	Traditional AI	Generative AI
Purpose	Recognize patterns, make predictions, and perform specific repetitive tasks.	Analyze data and inputs to create new patterns or content in addition to pattern recognition, predictions, or recommendations.
Industry Usage	Everyday AI to Improve productivity for routine or mundane tasks.	Game changing AI to bring in creativity and disrupt business models with new innovations.
Transparency	The decision-making process is more transparent and explainable since it is trained on limited data sets and predefined rules.	Explaining the results of a Generative AI system is complex since it is trained using complex algorithms on very large volumes of data.

(continued)

Table 1-1. (*continued*)

Aspects	Traditional AI	Generative AI
Strengths	Excels at prediction and optimization and produces more accurate results based on the inference derived from training data.	Produces creative and innovative outputs that may be biased and lack accuracy.
Weakness	Lacks cognitive flexibility and creativity as it operates on static rules and learned patterns. It cannot adapt to new situations without retraining or reason beyond what it was explicitly programmed or trained for. For example, a traditional AI model for fraud detection trained on past transaction patterns struggles to adapt quickly to new fraud tactics without retraining.	Generative AI outputs may have potential legal, ethical, political, ecological, social, and economic concerns. For example, it can hallucinate and produce factually incorrect results. This can mislead users.

Though traditional AI and Generative AI have distinct capabilities, they are not mutually exclusive. Traditional AI is more focused on improving everyday productivity, while Generative AI has the capability to bring in game-changing disruptions with its creative capabilities. The two can be used in tandem to build powerful solutions that were just a dream a few years back.

Key Takeaway: *Traditional AI is excellent for pattern recognition and prediction. Generative AI goes beyond: it creates new content—text, images, code, and more like humans.*

Benefits of Generative AI

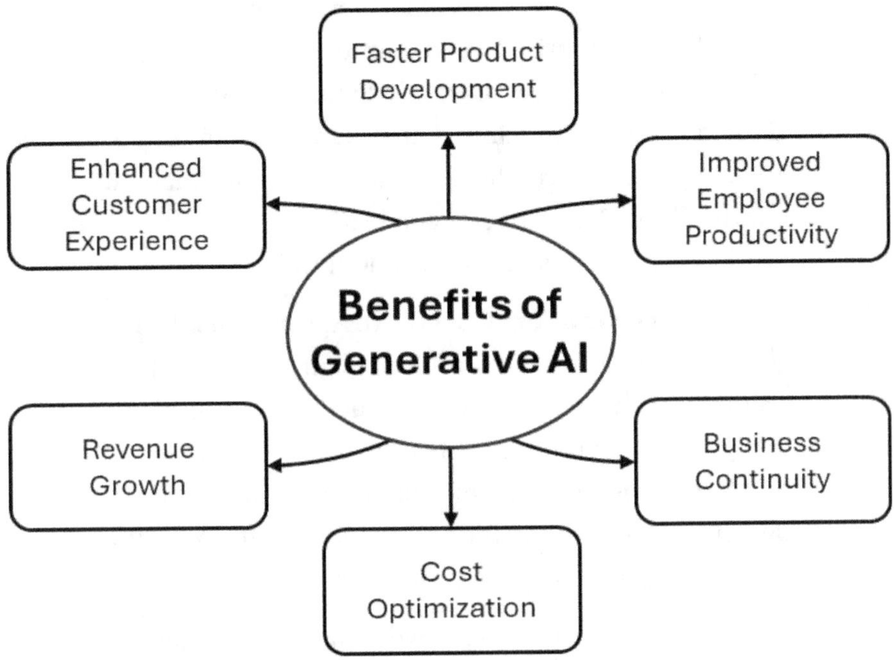

Figure 1-6. *Benefits of Generative AI*

Figure 1-6 shows the primary benefits of Generative AI, depending on the use cases where it is applied. The primary benefits are:

- **Customer Personalization**: Generative AI can analyze customer behavior and data and make highly customized recommendations aligning to their tastes and preferences at any time of the day. Generative AI-powered chatbots can provide 24/7 customer support and proactive advice while interacting in natural language. This enhances the overall customer experience.

- **Employee Productivity**: Generative AI can automate repetitive tasks and assist employees to perform their duties as a digital assistant. Generative AI can be used to summarize long documents, emails, or meeting notes. According to a Salesforce study, Generative AI cut the time to draft emails by around 30%. Applications can be automated to reduce manual efforts. Thus, Generative AI can lead to increased productivity and faster turnaround times.

- **Product Acceleration**: Generative AI can be used to get new and innovative product design ideas based on market trends, customer feedback, or competitive analysis. It can accelerate the process of creating wireframes, generating boilerplate codes, and even detect and fix bugs. All these capabilities can speed up the overall product development life cycle.

Some of the other objectives that are driving the adoption of Generative AI technologies across industries are as follows:

- **Revenue Growth**: Generative AI can identify growth opportunities and build new products and service offerings to grow businesses. It can reduce time to market for new product launches, leading to early monetization. Contextualization of ads and email campaigns can lead to better engagement and increased conversions.

- **Cost Optimization**: Generative AI can reduce manual efforts and increase efficiency for service delivery with automation of repetitive tasks and refined processes. It can reduce software development effort and cycles by generating boilerplate code, running test cases, and even help with debugging to fix bugs. It can help to reduce spending on marketing agencies and freelancers for marketing content generation.

- **Business Continuity**: Generative AI-powered chatbots can provide 24/7 customer and employee support while keeping support services running during crisis, holidays, and staff shortage. AI-driven workflows can keep core functions running without dependency on humans. Generative AI can also help to plan an unexpected future, risk management, and build resilient systems by helping to simulate "what-if" scenarios and be prepared with a contingency plan.

Applications of Generative AI

Generative AI can be put to use for various use cases across industries because of its ability to create different types of contents in response to inputs. Figure 1-7 shows the broad types of use cases where Generative AI can be applied.

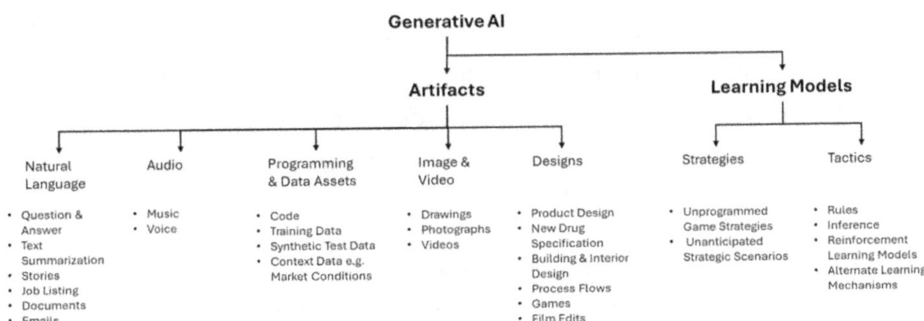

Figure 1-7. *Types of Generative AI Use Cases*

The most popular application of Generative AI technologies across industries are as follows:

- **Natural Language Processing**: Generative AI can be used for text classification and summarization, real-time language translation, named-entity recognition from text, document understanding and classification, and provide insights through sentiment analysis. The text generation and natural language processing abilities can be widely used for a variety of tasks, namely, literary compositions like stories, essays, poetry, creating social media content, summarizing documents, drafting letters, emails, and reports, creating job listings, answering questions and more - all using natural language.

- **Image and Video**: Generative AI can be used to create 2D and 3D images, avatars, videos, graphs, and more based on descriptions provided. Police and detective agencies can use it to create images of suspects and individuals based on imagery provided. Marketing agencies can easily create innovative images and videos for marketing to promote their products and services. It can be used to create realistic images for virtual or augmented reality, design logos, enhance or edit existing photographs,3D models for video games, and more.

- **Audio**: Generative AI for audio has various applications. It can be used for creating music compositions, remixing existing songs, and generating accompaniments for live performances. It can also be used to create voice-overs for audio books, educational videos, narrations for broadcasts and presentations, and marketing campaigns. Audio generation using Generative AI can be used to create accompanying noises for different video footage for greater sound effects.

- **Synthetic Data**: Data is needed to train AI models and simulate real world scenarios when building autonomous systems like self-driving cars or robotic applications or medical diagnostic devices. However, the data to test applications or systems may not always be available because of compliance, privacy or security requirements or organizational silos or that data simply does not exist. Generative AI brings in a solution to address this situation with its capability to generate synthetic data that mimics real world scenarios without compromising privacy or security. Generative AI can enhance existing training datasets. It can replace real-world sensitive data with synthetically generated data that can be used in healthcare, financial, and other regulated industries without violating regulatory requirements or raising privacy concerns. Generated synthetic data can simulate real-world environments, such as weather conditions, traffic patterns, or market fluctuations, for testing autonomous systems, robotics, and predictive models without real-world consequences.

- **Programming**: Today, Generative AI tools have the capability to generate efficient code in different languages like Python, Java, C#/.Net, JavaScript, Go, Ruby, and many more based on conversational prompts. Code can be generated based on general best programming practices, organizational governance, or natural language description of designed code. It can also assist in code completion and provide code explanation. Hence, professional programmers can use Generative AI tools to generate their code, fix bugs, or suggest alternative methods to make their code faster and more efficient.

- **Design**: Generative AI is poised to redefine how we ideate, prototype, and develop innovative products that captivate users. It enables product designers to explore more ideas and product experiences. With a simple prompt, Generative AI can unlock creative possibilities, inspiring teams to think beyond the boundaries of traditional ideation. It can serve as a catalyst and shorten the physical product design life cycles significantly, faster than the traditional methods. Generative AI can reveal untapped market opportunities and suggest product features, customer needs and preferences that were earlier overlooked. It can generate novel prototypes, concept images, mood boards, and storyboards that can inspire bolder exploration and bring forward first-of-their-kind ideas. Once designed, it can also accelerate the phases of concept refinement and testing. Overall, Generative AI can be used for new product design, drug design, building and interior design, process flow design, and much more.

Key Takeaway: *Think beyond automation—Generative AI unlocks net new capabilities previously not possible with traditional AI.*

Modalities of Generative AI

Modality is the type and form of output that a Generative AI application can generate. Generative AI can produce novel outputs that are very similar and appear to be generated by humans. It can produce outputs in six key modalities. This enables Generative AI to produce outputs in formats beyond text, providing rich digital experiences. Figure 1-8 shows the modalities of Generative AI.

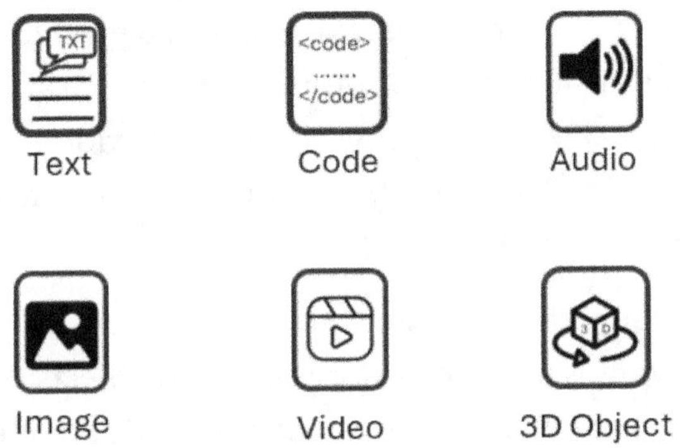

Figure 1-8. *Modalities of Generative AI*

- **Text**: These AI-generated, coherent, and meaningful written content resembles natural human communications. Examples include summarizing documents, writing customer-facing materials, generating articles and blog posts, or explaining complex topics in natural language.

- **Code**: Generative AI can generate computer code in a variety of programming languages. It can provide explanations of the generated code and can help debug issues in code. Examples include generating code from natural language description or auto complete code-snippets or translating code between programming languages.

- **Audio**: Generative AI can generate new lifelike audio in natural, conversational, and colloquial styles in different languages and tones. Examples include text-to-speech, voice cloning, dubbing voice in movies and videos for translation, AI-powered call centers.

- **Image**: Generative AI can generate original and realistic visuals based on textual descriptions provided in natural language or visual prompts. Examples include creating a visual image of how a product would look in a customer's environment or reconstructing a suspect image based on descriptions provided. DALL-E can create high-quality 2D product mock-ups from text prompts.

- **Video**: Generative AI can take user prompts and create videos, with scenes, people, and objects that are entirely fictitious. Videos can be a realistic animation of people or objects, or it can be created based on existing footage. Examples include creating a video to demonstrate a new product or simulating dangerous scenarios for safety training.

- **3D Objects**: Generative AI can generate data representing intricate and innovative 3D objects from text or images with precision and speed. This simplifies the modeling process and allows you to explore a wide range of design possibilities. Examples include creating a 3-D view of the structure of a building or a product or a molecule. Game designers can use this capability to quickly create realistic animation of characters and environments for their games. Furniture manufacturers can create 3D models of sofas and chairs from text prompts.

Key Takeaway: *Be creative to unlock the possibilities of Generative AI that can process inputs and produce outputs in different forms, namely, text, code, image, video, audio, and even 3D objects.*

How Generative AI works

Generative AI starts with an input called "prompt." The input can be in the form of text, image, video, audio, or any input that the AI system can process. The input is analyzed by the system to understand the task and the specific requirement details, which is then fed to Generative AI foundation models.

These foundation models are fed and pre-trained with a large dataset of existing text, image, and videos using unsupervised and semi-supervised learning approaches. Neural networks and deep learning algorithms help to identify patterns and structure in the training data. During training, the neural network adjusts the connections (weights and biases) between its neurons to recognize and learn the underlying patterns, structure, and relationship between the data. The patterns learnt from the training data are used to create similar but original content that is new and previously unseen in the training data.

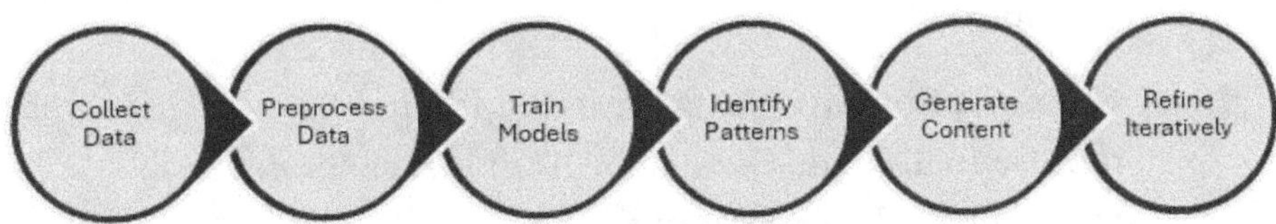

Figure 1-9. *Generative AI Life Cycle: From Training to Inference*

Figure 1-9 shows the key steps involved in the process used by Generative AI to generate responses for different use cases as described below:

- **Collect Data**: Large volumes of example dataset is collected to train the Generative AI models

- **Preprocess Data**: The collected data is pre-processed to remove noise and errors before it is fed to the model.

- **Train Models**: The Generative AI models analyze the training dataset to learn and understand the patterns, structure, relationships, and underlying rules governing the content. The training process teaches the model to recognize how different variables are related to each other in combination.

- **Identify Patterns**: A trained Generative AI model converts what it learns from data into a compact internal summary, called an **embedding.** The embedding captures the most important patterns and relationships and helps the model quickly understand context and generate relevant outputs. For example, an AI system is trained on thousands of customer support articles. It converts this learning into **embeddings** that capture the meaning of each article. When a customer asks a question, the AI compares the question's embedding with the stored ones to quickly find the most relevant information and generate an accurate response.

- **Generate Content**: Using the embedding representation, the Generative AI model generates new content based on user input.

- **Refine Iteratively**: The Generative AI model is refined with new learnings iteratively for better results. Hence, the generated output is evaluated, and the areas of improvement are identified. The feedback is then used to adjust the model's parameters to improve the quality and realism of the generated content. This process continues until the model produces satisfactory results.

Generative AI models differ based on the training process and the approach to generate output. The next section discusses the different types of popular Generative AI models as shown in Figure 1-10.

Key Takeaway: *Generative AI models are trained on massive volumes of data to learn patterns and generate new content in response to prompts that guide the generation process. The quality and relevance of outputs improve with iterative training.*

Key Generative AI Models

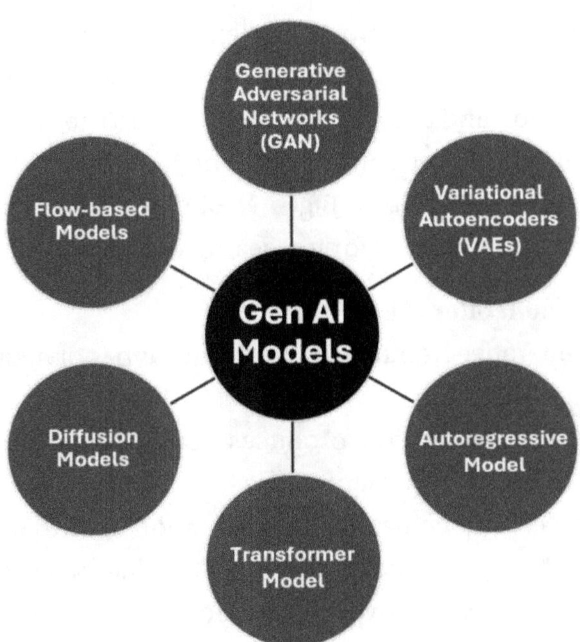

Figure 1-10. *Types of Generative AI Models*

Generative Adversarial Networks (GANs)

A Generative Adversarial Network is made up of two neural networks—the **generator** and the **discriminator**.

These two neural networks compete against each other. The generator tries to create new synthetic data (e.g., image, text, or audio) from random noise. The discriminator tries to discriminate between real data (in a training data set) from the generated data. The generator aims to create increasingly realistic data to deceive the discriminator, while the discriminator improves its ability to differentiate real data from generated data. Through this competition and adversarial process, both generator and discriminator reach an equilibrium, where the generator creates increasingly realistic data samples, and the discriminator has a harder time distinguishing between real and fake data. This enables GAN to generate highly realistic images.

Generative Adversarial Network models have been successfully used in image synthesis, art creation, video generation, deep fake creation, and super resolution and data augmentation tasks. They have also been used in the field of bio-science to generate high quality synthetic data for research and analysis. StyleGAN, BigGAN, SRGAN, AttnGAN are some examples of GAN models used for image processing.

Variational Autoencoders (VAEs)

It is a type of Generative AI model that uses two types of neural networks—an **encoder** and a **decoder**.

The encoder neural network produces a compressed and simplified version of the input data represented in a lower dimension latent space. The latent space is a mathematical representation of the data. It can be thought of as a unique code representing the data based on all its attributes. For example, if studying faces, the latent space contains numbers representing eye shape, nose shape, cheekbones, and ears. The decoder reconstructs the data from the latent space that resembles the original input.

Variational Autoencoder (VAEs) models can be used for realistic data/image generation, data compression, anomaly detection, and drug discovery.

Autoregressive Models

Autoregressive models are a class of machine learning models that predict the next element in a sequence of data using a probability distribution, taking into account the previous inputs.

Autoregression is a statistical technique used in time-series analysis that assumes the current value of a time series is a function of its past values. These models use the same mathematical technique to generate elements sequentially, predicting each one based on the preceding ones.

Autoregressive models are commonly used for natural language processing to respond in ways that humans can comprehend, text generation, image synthesis to sharpen and reconstruct images, time series predictions for forecasting stock prices, weather, and traffic conditions, music generation, and data augmentation to generate synthetic training data.

Transformer Model

Transformer Model is a type of neural network-based deep learning model that uses a mechanism called **self-attention** to determine the importance of different parts of input data.

The transformer model analyzes data sequentially and learns the context and the meaning by tracking relationships in sequential data. Self-attention techniques are used to understand the relationship between pairs of tokens (words) in a sentence. This helps to track the relationship even between distant words in a sentence or paragraph and how they are related to each other. Transformer models process input sequences in parallel, which makes them faster and more efficient when compared to other models used for sequential data analysis like Recurrent Neural Network (RNN) and Convolution Neural Network (CNN).

Transformer models have been widely used for Natural Language Processing (NLP) for text summarization, language translation, question answering, and various other text generation capabilities. It can also be used to generate images and music, detect anomalies, and prevent fraud, programming, and more. ChatGPT, Copilot uses transformer models to understand long conversations, maintain context across multiple questions, and generate coherent human-like responses. Transformers power today's most impactful AI systems by understanding context, relationships, and sequences at scale—making them ideal for language, vision, recommendations, and decision support. Banking chatbots are built using transformer models to answer policy questions. A more detailed working of the model will be covered in a later chapter in this book.

Diffusion Models

A diffusion model is a type of Generative AI used to create images, audio, and videos. It's based on the concept of **diffusion** in physics, where molecules move from areas of high concentration to low concentration, like how a drop of food coloring spreads in water until it evenly colors the liquid. In machine learning, this idea translates to adding random noise to data (like an image) until it becomes noisy.

The model works by first training a neural network to gradually destroy the data by adding noise. Then, it learns to reverse the process and recover the original data from the noise. After training, the model can start with random noise and generate high-quality, structured outputs by reversing the noise. Diffusion model comes with benefits of scalability and parallelization, that makes the training process highly efficient. They generate high-quality images with fine details and realistic textures.

This method allows diffusion models to create highly detailed images. Popular examples include DALL-E 2, Midjourney, and Stable Diffusion, which generate realistic images from text descriptions. Diffusion models can also be used for image-to-image translation while preserving the semantic information and visual coherence. Newer models like Stable Diffusion 3, launched in 2025, enhance image generation. It can also be used for image search.

Flow-based Models

Flow-based models are a type of deep generative model that uses
a sequence of reversible mathematical steps called **invertible
transformations** to model complex data distribution. A transformation
model is said to be "invertible" if it can transform an input to output and
apply the inverse function on the output to get back the input.

Flow-based models create designs by mapping data patterns—great
for prototyping. They map samples from the input distribution to samples
from the target distribution. By learning the distribution of data it is
trained on, the model generates new samples similar to the training data
by applying the transformations it has learnt. Because of the reversibility,
flow-based models can move back and forth between simple and complex
distributions.

RealNVP (Real-valued Non-Volume Preserving), Glow, FFJORD (Free-
form Jacobian of Reversible Dynamics), SRFlow (Super Resolution Flow),
Flow++, and MAF (Masked Autoregressive Flow) are some prominent
examples of Flow-based generative models. Flow-based models can be
used to generate high-fidelity images, synthesize realistic speech, generate
sequence of video frames, estimate probability density of a given dataset,
model new molecular structures, solve inverse problems, accelerate or
replace computationally expensive simulations in physics and many more.

Key Takeaway: *There are many Generative AI models. Know the model
that suits the business use case—not all models are fit for all purposes.*

Key Patterns of Generative AI Usage

Generative AI has the potential to disrupt a wide range of industries by
a variety of innovative use cases. It can be used for content creation to
generate new contents such as text, image, video, audio, or a combination
of these. The generative capability can be used to produce output in the
same mode as the input or in a completely different mode. The output can
also be in a combination of different modes as shown in Figure 1-11.

For example, a Generative AI model can take the chat transcript and create a summary of it. In this case, both the input and output are in text format. It can also generate a description for a given image or take an input text description and create a corresponding image output. This is a cross-modal Generative AI capability. Taking it further, a multi-modal Generative AI capability can take a description of input and generate a video with audio.

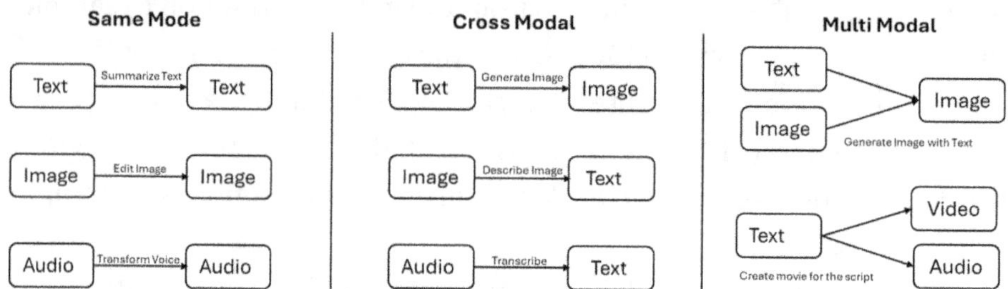

Figure 1-11. *Modalities of Generative AI*

Let's now look at the different uses of Generative AI in the industry for various use cases.

Text Generation

Generative AI can take a text input as a structured data, code, or transcription and generate a completely new output in text format. The text generation capabilities can be used for the following operations:

- Text completion

- Text classification

- Text summarization

- Text translation

- Text correction

- Text manipulation

- Question and answering

- Sentiment analysis

- Simple analytics

- Style translation

In an enterprise, these operations can be used for the following and many more use cases:

- **Content curation**: Curate content from thousands of scanned documents with summarized information of relevant industry trends and regulatory changes.

- **Enterprise search**: Search across internal documents, emails, and knowledge bases and returns a clear, summarized answer instead of a list of files.

- **Marketing content creation**: Create personalized campaign copy, social media posts for different customer segments.

- **Personalized sales**: Create customized sales proposals and pitch decks by pulling relevant information from product catalogs, pricing, and past deals.

- **Research report generation**: Analyzes data and documents to produce a concise research report with key insights and recommendations.

- **Customer service and support**: Provide answers to customer questions in real time using information from support knowledge bases.

- **Meeting summarization**: Convert meeting conversations into concise summaries with key decisions, action items, and next step.

Morgan Stanley has built a chatbot for their financial advisors that uses Generative AI models trained on over 100,000 curated research documents. It provides expert advice for their financial advisors with clear answers with reasoning and reference sources.

Code Generation

Generative AI can be used for software coding. It can generate code, provide explanations for code, and even find bugs or defects in code. It can be used to write code to implement specific business logic or create unit test cases for a developed code. Copilots embedded within code editors can suggest single lines or entire blocks of code in real time based on user comments describing the logic. Generative AI can even translate code from one programming language to another, which can be used for legacy code modernization like transforming COBOL code to java.

Image Generation

Generative AI technologies can be used to create realistic images based on the input description provided. The description can specify a subject, a location, environment, or style to use for the image generation. It can also be used to enhance existing images with lighting, color, or style, or restore old images. The image enhancement capabilities can be used to generate realistic images from a sketch or semantic image. Midjourney, DALL-E can produce unique and eye-catching images for social media posts. Nutella used Generative AI to create unique designs for its jar labels.

Video Generation

Generative AI technologies can be used for creating high quality video content like training videos with diverse avatars and voices. It can automate time-consuming tasks like creating new videos with special effects and animation. Generative AI can be used to create video clips for social media posts from text prompts. It can be used for product visualization under different environments and settings without the need for expensive video shoots. Network Rail created 600 videos for their training purpose in five months using Generative AI technologies. It helped them to cut down the time to produce video by 95% and quickly adapt the video to the changing needs.

Audio Generation

Generative AI can be used to compose new music from patterns and styles of music the model is trained on. Content creators are creating royalty-free background music tailored to the mood of their videos. It can also be used to produce high quality realistic speech audios from text inputs in natural language. YouTubers are using various tools that use Generative AI technology to produce clear and consistent voice-overs for their audio and video content. Generative AI is being used by businesses to generate voice for their internal training content, product demo videos, promotional ads and radio spots. It can even be used for speech-to-speech conversion to support voice cloning and dubbing in movies.

Designing New Things

Generative AI capabilities can be used to create new designs. Below are some examples where Generative AI can be used to create new designs:

- Design new building plans, stunning elevations, and aesthetic interiors

- Design lightweight and aerodynamic automobiles and aircraft

- Design components of automobiles and machineries

- Design new chips for electronics and computers

- Design new molecules for compounds and drugs

IBM used Generative AI to identify a new molecule in a few months, which otherwise would have taken ten years, costing 10 to 100 million dollars. This was 100 times faster than traditional methods. Similarly, Insilico used Generative AI to create a new drug for a rare disease named

Pulmonary Fibrosis. This reduced the time from start to phase 1 clinical trial to 30 months, which otherwise would generally take anywhere between three to six years.

Of all the possibilities that Generative AI brings, text, image, and video generation have a high technical feasibility and are the most used capabilities in most business use cases. Designing new things needs a lot of research and past data of high quality for training the Generative AI models, since such data may not be easily available. Hence drugs, material, or chip design have a lower feasibility, though they can have high business values.

Key Takeaway: *Generative AI supports both intra-modal and cross-modal generation, which makes it highly flexible for different business needs. For example, it can turn product descriptions into videos with voice-overs.*

The Promise of Generative AI for Business Innovation

Generative AI is a technology catalyst for innovation with the potential to make significant and disruptive changes across industries. It promises to unlock the next wave of productivity by enabling customized customer experience and generating creative content at scale.

With its ability to quickly generate new ideas and design options, Generative AI has opened up a world of possibilities for businesses looking for innovation. Businesses can tap into the creative potential to generate new ideas, designs, and solutions, augment human creativity, and provide a competitive advantage.

Generative AI is already being used to design new buildings, apparels, or consumer products. In the healthcare industry, it can analyze medical data like MRI scans and X-rays with accuracy and speed, identify patterns and correlations in patient records, and provide more accurate diagnosis and personalized treatment plans.

Automotive powerhouse Tesla uses Generative AI to optimize its production workflows and improve efficiency of its manufacturing process. Beyond that, it can automate repetitive and mundane tasks and free human resources to focus more on creative and strategic activities. It can supplement human creativity and also produce and identify novel ideas. It can evaluate and refine ideas to improve the quality of raw ideas. Generative AI brings in the opportunity for private and government organizations to augment human creativity and democratize innovation.

Generative AI Usage Across Business Functions

Generative AI is going to transform roles and boost performance across several functions in the industry. However, the ones that are going to see a significant impact are marketing and sales, product research and development, customer service operations, and software development. It is estimated that Generative AI will boost worker productivity across various functions by 35 to 70%. New use cases unveiled by the generative capability can drive revenue growth by 15 to 40%. As Generative AI technology develops and matures, it has the potential to unlock completely new frontiers of creativity and innovation.

Table 1-2 lists the business functions that will see moderate to significant impact due to Generative AI.

Table 1-2. *Impact of Generative AI Across Business Functions*

Business Functions with High Impact	Business Functions with Moderate Impact
• Sales and Marketing • Customer Service Operations • Product Research and Development • Software Engineering	• Supply Chain • Manufacturing • Finance • Risk and Compliance • Talent and Organization • Legal • Procurement • Strategy and Pricing

Some of the top use cases of Generative AI across different business functions in industries are as follows:

Marketing and Sales

Generative AI can have a significant positive impact on the marketing and sales functions of any organization because of natural language communication and personalization ability. It can be used in marketing for content creation, drafting emails, social media posts, blog articles, creating audios, and videos for personalized campaigns. It can analyze marketing trends and customer data from various sources like social media, news, research, product information, and customer feedback, allowing it to develop a highly tailored and effective marketing and sales strategy. Generative AI can generate campaigns that are tailored to the segment, language, interest, and demographics of the customer.

Additionally, it can drive sales with dynamic recommendations that influence customer purchase decisions. Generative AI can identify and prioritize sales leads by creating a comprehensive customer profile and making recommendations to staff on actions to increase client engagement. It can provide information about client preferences, which can be considered during the sales process to increase the probability of sales. AI-powered chatbots can automate and improve the lead qualification by engaging with website visitors, asking them relevant qualifying questions and directing the leads to the right sales representatives. Overall, Generative AI can be used to increase the probability and improve the experience and productivity of sales.

Customer Service Operations

Customers can directly interact with Generative AI-powered chatbots in their natural language to get answers to their queries any time regardless of their language or location. The ease of conversing in natural language and personalized responses to complex queries improves the overall interaction experience. Generative AI can also provide real-time assistance and tailored suggestions to customer service agents during their phone conversations with customers. AI agents can also record the summary of the interaction with customers and the actions taken. Generative AI can thus improve the customer experience and agent productivity through digital self-service and augmenting agent skills. It can not only help to resolve customer queries and issues during their initial contact but also reduce the overall response and resolution times.

Delta Airlines has used Generative AI to transform customer experiences while improving operational efficiency. Companies are using chatbots trained in specific product knowledge for customer support and provide product recommendations. It is also used to create qualified sales leads, automate email responses, and analyze customer feedback. Companies can cut their customer service costs by up to 30% using Generative AI.

Product Research and Development

Generative AI can be used to understand the market and customer needs, draft ideas, and product solutions. It can create virtual designs and simulations of the product and plan its physical testing. This can reduce the overall R&D cost. Life sciences companies are using Generative AI foundation models in their R&D to generate candidate molecules, accelerating the process of developing new drugs.

Generative AI can also suggest improvements in existing designs to address customer feedback. It can also help to optimize design and suggest materials that can not only improve product efficiency but also reduce overall cost for production and logistics.

Software Engineering

Software engineers can use Generative AI as their partner in pair programming. Generative AI can not only generate boilerplate code, but it can also explain the logic implemented, suggest code refactoring following best practices, debug and find issues in code, and even help to optimize the code for better performance. Generative AI can also be used for creating unit test cases with better code coverage. It can automate and optimize testing to highlight potential problems. It can also generate diverse test data for different scenarios to maximize coverage. All of this can significantly improve productivity for software development by 20 to 45%. This gives more time for software engineers to focus more on better architecture and design, in turn improving their satisfaction and overall work experience.

Key Takeaway: *Generative AI can personalize outreach, create intelligent agents, design products, and code software. It can deliver both cost savings and revenue uplift, depending on the function and implementation maturity.*

Generative AI Is a Game Changer for All Major Industries

Almost all industries have started exploring the use of Generative AI. The maximum usage of Generative AI is for productivity and efficiency improvements, followed by improved and personalized customer service experience. Other benefits include improved revenue and better risk management. Here is a brief summary of use cases across different industries.

Banking and Financial Services

In the banking industry, Generative AI can be used for risk assessment and real-time fraud detection, ensuring regulatory compliance and providing personalized customer experiences. It can also help to calculate the creditworthiness of an individual or entity based on their income, employment, and credit history, thus helping banks and financial institutions in more informed decision-making in lending and investment processes. JP Morgan Chase is pioneering the development of gen AI with index GPT. It helps to get deeper insights into different investment options and helps to optimize investment strategies.

Capital Markets

Generative AI is a perfect fit to analyze market data and trends, news events, social media, and historical data at lightning speed and make decisions for the right trade investments. It can be used to gauge the market sentiments and execute trades with precision and efficiency that benefits the investors. Investment and brokerage firms can use Generative AI to generate earnings summaries, market commentary, and investment research notes in real time. Asset Management companies can predict customer churn using Generative AI. Other use cases of Generative AI in the capital markets include chatbots for customer queries, interpretation of complex regulatory changes and automate drafting of compliance documents, and generating alerts from flagged transactions and communications.

Retail and Consumer Goods

Generative AI can be used in the retail and consumer goods industry to provide personalized shopping experience with hyper-personalized product recommendations, predict future demands from historical data, and help retailers optimize their inventory levels. It can also be used for dynamic product pricing based on market trends, competitor pricing, and customer behavior. Generative AI is all set to redefine the shopping experience, making it personalized, efficient, and more engaging. Shopping experience is going to get an uplift with Generative AI with its ability to support visual search capabilities. Visual search will make it easy, convenient, and faster for customers to find the right product by analyzing the patterns in photos or descriptions of the product. Customers can simply share the picture or description of the product they are looking for, and Generative AI can help to find the same or similar product from the catalog. Generative AI can not only optimize business operations but also provide a shopping experience that resonates with customers, fortifying loyalty and driving business growth.

Healthcare and Life Sciences

In healthcare, Generative AI can be used to enhance medical images like X-rays and MRIs, diagnose patient ailments and even recommend personalized treatment plans. It can also build patient information summaries, create transcripts of verbally recorded notes, or even find essential information in medical records more efficiently than humans.

In the Life Sciences industry, Generative AI can be used to accelerate new drug discovery and development and act as a catalyst for disruptive medical advancements.

Travel and Hospitality

Generative AI can provide real-time updates and recommendations to travelers to help them plan their itinerary and travel better. It can provide a personalized travel itinerary to travelers based on their preferences and budget. Travel companies can build Generative AI powered chatbots that can assist customers with their travel bookings by answering their queries that they may have during the booking process. Generative AI can be used to craft personalized messages, finely tuned to increase customer engagement and build trust that plays a crucial role in acquiring new customers and fostering loyalty. Language translation and localization capabilities can break the language barrier and help travelers to explore foreign countries that do not speak their native language. Generative AI can also provide personalized packing recommendations and risk assessments for travel destinations, routes, and activities. Travel companies can also benefit from dynamic pricing based on seasonality, competitor pricing, and traveler preferences with insights from historical data. By adopting Generative AI, travel companies can increase customer engagement, streamline processes, and offer tailored recommendations that can increase customer satisfaction and loyalty.

Media and Entertainment

Generative AI is reshaping the media and entertainment industry. It is poised to bring in a new era of personalized and immersive experiences. Generative AI will impact the entire value chain of the media and entertainment industry—from how content is produced, marketed, and monetized and, later scheduled and distributed optimally to specific targeted audiences, keeping them engaged. It can make personalized recommendations for viewers based on their preference, viewing history, and current trends. Tailored content is going to drive the satisfaction, engagement, and loyalty of the audience. Generative AI can be used to localize content for different languages, regions, and cultures. Magellan TV uses Generative AI to internationalize its gaming industry, uses Generative AI to create in-game virtual items, such as costumes and gear, reducing the manual efforts of the game designers. Such new and unique items help to keep the games fresh and exciting for the players, improving the overall gaming experience. For entertainment, Generative AI is being used for various creative endeavors like music composition, video generation, and virtual reality-based games. There are many ready-made tools and frameworks available that use Generative AI to create music remixes, special effects on videos, or even generate new games based on input prompts. Digital avatars powered by Generative AI can be used as a cost-effective and efficient replacement for real professionals. Coca Cola has used DALLE-2 to create images and videos for its marketing campaigns. Personalization, improved production efficiency, targeted marketing and promotions, better audience experience, and reduced cost through AI-powered automation are some of the main advantages that the media and entertainment industry can look for from Generative AI usage.

Education

Generative AI has the potential to completely transform the learning and education experience. It can help with personalized learning, course design, content creation, and streamlined the evaluation process with automated assessment and grading. It can also be used to create a virtual tutoring environment where students can interact and learn from virtual tutors and receive real-time feedback. Additionally, the natural language translation capability of Generative AI can provide a more interactive and engaging learning experience.

Automotive Manufacturing

The Automotive industry is using Generative AI to create more energy-efficient and high-speed cars. Key use cases include automotive design and prototyping, manufacturing process enhancements, quality control, preventive maintenance, and supply chain optimization. Generative AI can also enhance driving experience with personal voice assistants, personalized in-car infotainment, and intelligent routing and navigation. Proactive preventive maintenance using Generative AI can help to significantly reduce breakdowns and lower maintenance costs. Generative AI can also be extensively used to simulate environments and scenarios used for the testing and validation of autonomous cars. The benefits include better safety and fuel economy, a more personalized and convenient driving experience, and smarter traffic management.

Energy and Utilities

Generative AI in the energy and utilities industry can drive efficiency, reduce costs, optimize operations, and enhance safety. It can help companies optimize drilling processes and predict equipment maintenance needs. They can make data-driven decisions to optimize cost and improve sustainability.

Some of the use cases where Generative AI can be used in the energy and utility sector include reservoir simulation and modeling, drilling optimization, production forecasting, seismic imaging and interpretation, environmental impact assessment, exploration target identification, carbon capture and emission reduction, and refinery process optimization. Generative AI can also provide immediate technical support through conversational agents, improving efficiency, reducing costs, and minimizing downtime. It has also revolutionized energy trading with real-time insights and streamlines reconciliation that has helped to reduce financial risks.

Key Takeaway: *Every industry has a Generative AI entry point. Identify the right one based on business priorities and data readiness.*

Conclusion

Generative AI comes with immense potential to redefine the future of business. It is the driving force for innovation, operational excellence, and transformative changes across all industries. It can be used to craft exceptional customer experiences and unlock completely new business opportunities.

But like any other technology, Generative AI too has many risks associated within that can be misused, and downsides that cannot be overlooked. Hackers can craft more convincing phishing scams and malware that may be hard to detect. The fear of losing their job to AI has been bothering many. Lots of concerns have been raised about intellectual property rights and bias in AI outputs. Hallucinating due to incorrect information can break human trust in these systems. Hence, it is paramount to maintain vigilance over how these AI systems are developed, deployed, and used. To responsibly harness the full potential of Generative AI, organizations must have a comprehensive strategy that integrates technology with cultural readiness. Risks and concerns with Generative AI must be addressed at the foundation. For example, bias can be mitigated with diverse training data and regular audits. It is critical to build trust by implementing a robust AI governance framework and practices that prioritize transparency, fairness, and accountability. Only then can businesses unlock Generative AI's benefits and protect its stakeholders. To lead in the age of Generative AI, executives must act now by identifying high-impact use cases, piloting responsibly, and building cross-functional GenAI task forces. Those who wait risk falling behind.

CHAPTER 2

A Double-Edged Sword: Benefits and Risks of Generative AI

Introduction

Generative AI is rapidly transforming industries with its rich capabilities to generate human-like text, images, media, creative solutions, and ideas. While this undoubtedly comes with lots of advantages and benefits, it has its dark side too. While it can be used for innovation and creativity, the same creative capability can be misused by malicious actors with a wrong intention. Generative AI can pose several serious risks—it can spread fake news and misinformation, amplify societal bias, weaken cybersecurity, and even make people overly dependent on machines, reducing human creativity. When in the wrong hands, it can be misused for manipulation, fraud, bullying, and harassment.

It is our responsibility to ensure that Generative AI remains a tool for good and is used for the betterment of society. This needs an ethical framework with a holistic governance, complete transparency and strict regulations to ensure that this powerful technology has a positive impact on human society.

B. De, *Generative AI for Business Innovation*, https://doi.org/10.1007/979-8-8688-2682-5_2

This chapter takes a thoughtful look at both the benefits and the risks associated with Generative AI. It provides real-world examples of misuse and offers practical guidance on how to use this powerful technology responsibly. For instance, deepfakes have been used to scam businesses, like a $25M fraud in Hong Kong. The chapter covers frameworks for ethical AI deployment, and actionable governance strategies to help organizations build Generative AI applications that benefit mankind. For business leaders, this means more than just keeping up—it means leading the way by setting clear boundaries, focusing on low-risk, high-value use cases, and building a workplace culture that values ethical innovation.

Promise of Generative AI: Business Benefits

Generative AI use cases are making their way from pilot to production. More and more organizations are increasing their Generative AI investments and using it across multiple of their business units. This trend is fast growing across firms in all industry domains. Generative AI is driving value across the organization. It is primarily used for customer services, marketing, and sales business functions. IT functions are using Generative AI for software development, infrastructure, and operations. Even finance, HR, procurement, and legal are not far behind in their adoption of Generative AI. Organizations are using Generative AI not merely to automate tasks - but fundamentally to transform how work is performed.

Figure 2-1 shows some of the benefits that businesses can expect from the adoption of Generative AI.

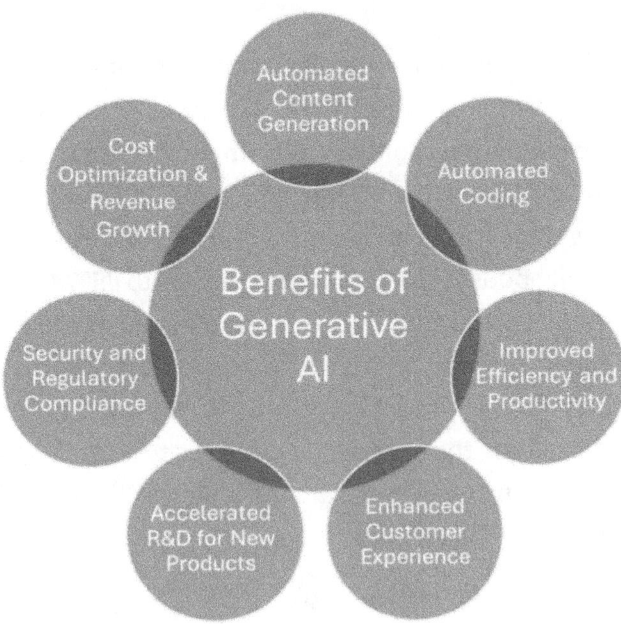

Figure 2-1. *Benefits of Generative AI*

Automated Content Generation

Creating great content can be tedious and time-consuming. Fortunately, Generative AI streamlines the process by quickly generating engaging product descriptions, blog posts, and articles. It can efficiently craft attractive social media posts for marketing and branding. AI can also generate personalized messages tailored to individual customers' interests and preferences, enhancing engagement. It can also assist with email responses or large-scale campaign content. By automating content creation, Generative AI not only boosts productivity but also increases customer attraction, conversion, and retention—delivering scalable results for businesses.

Automated Coding

Automated code generation to write new code is another great use case for Generative AI. Anthropic's Claude.AI, GitHub Copilot, and OpenAI's ChatGPT have been helping software developers write code at almost half the time, significantly boosting their productivity.

Software engineers can use Generative AI as a pair-programmer that can potentially boost their productivity by up to 45%. Generative AI can generate initial code drafts, correct and refactor code, explain code segments, debug errors, and analyze root causes. It is predicted that in the near future, entry-level coding tasks may be increasingly automated using Generative AI—pushing the skill sets needed for software engineering toward architecture and design. Newer tools like xAI's Grok 3, launched in 2025, enhance coding efficiency.

Improved Efficiency and Productivity

Generative AI can help to free up time by automating a lot of manual steps. It can automate repetitive tasks, enabling teams to focus on work of higher value. Enterprisees are already seeing early results with productivity gains using Generative AI in key areas like marketing, IT support, HR, software development, and customer service. The list is long and is only limited by our imagination.

Generative AI can be used by legal professionals to review and draft legal documents. It can identify relevant clauses, potential risks, discrepancies, and streamline contract reviews. In the banking and financial sectors, Generative AI can understand and respond to customer queries in natural language to streamline many processes. It can check account balances and transaction history, transfer funds, and also provide financial insights. In the manufacturing industry, companies like Siemens are using Generative AI to optimize the design of turbine blades, cutting prototype time by 30%. Generative AI can accelerate manual and repetitive tasks and efficiently generate new content, freeing up human resources for more strategic work. According to EY and HR Dive, the productivity gain from Generative AI is estimated to be between 20 to 40% without any loss in quality, depending on the roles and implementation.

Enhanced Customer Experience

Generative AI can be used to provide personalized customer experience, which is key to engaging and retaining customers. This is one of the key benefits of Generative AI. It can analyze vast amounts of customer data and provide tailored recommendations, marketing messages, or shopping experiences for individuals. Digital self-service for customers by interacting with Generative AI-powered chatbots in natural language is a totally new experience. Customers can get immediate personalized responses to their queries from Generative AI-powered chatbots. This improves customer engagement and satisfaction, builds brand loyalty, and ultimately drives sales and revenue.

Accelerate R&D for New Products

Generative AI can empower designers and innovators with a wide range of design possibilities to explore. It can be used for architectural designs and developing product prototypes that can spark innovations. Generative design can help with rapid ideation and create new attractive products and services of higher quality. It can optimize designs for manufacturing that can reduce logistics and production costs, helping businesses to stay ahead of their competitors.

Foundation models can generate candidate molecules and accelerate gene sequencing and the process of drug development in the life sciences industry. Insilico Medicine used Generative AI to accelerate their drug discovery process and developed new therapies. It used Generative AI to discover the target drug and design the molecule for the treatment of idiopathic pulmonary fibrosis, a relatively rare respiratory disease that causes progressive decline in lung function. The drug discovery process went from novel target discovery to Phase 1 in-human trials in under 30 months, which is about half the time the process would take with a traditional approach and at a fraction of cost.

Security and Regulatory Compliance

Generative AI can be used to analyze code for any security vulnerabilities and identify unused or malicious code. AI-powered tools can analyze vast amounts of data to detect anomalies and identify security threats in real time, enabling businesses to proactively take action to mitigate risks and protect their assets and reputation. Generative AI models can flag anomalies in user and network behavior like unusual login times or network traffic. It can also analyze security logs and identify suspicious activities and simulate cyberattacks to find vulnerabilities before hackers can find them.

Generative AI can also check configuration against compliance standards and suggest fixes. It can also track changes in laws and regulations and automatically update compliance policies.

Cost Optimization and Revenue Growth

Generative AI, through its multifaceted solutions, can drive cost reduction and savings. The automated content generation capabilities will significantly reduce the reliance on manual labor. Generative AI can save time and optimize the cost of doing manual, repetitive, and time-consuming tasks.

The launch of new products and services powered by Generative AI, will open up new revenue channels for the business. In the long term, it will have a significant impact on the top-line revenue growth. Improved customer satisfaction and reduced customer churn will result in more sales and boost revenue. With its ability to analyze customer data and behavior, Generative AI can also be used by business to cross-sell and upsell their products.

Hidden Risks of Generative AI

As Generative AI adoption accelerates, organizations must understand and manage the risks that come with its benefits. Businesses are looking at innovative ways to use Generative AI to interact efficiently with their customers and drive business growth. However, like any other new technology, Generative AI is not without its downside. It has many risks and concerns associated with it that need to be addressed to make it trustworthy. If Generative AI does not give the correct result, the implications can be significant or even fatal, depending on the use case. For example, an incorrect diagnosis or treatment recommendation can lead to death of a patient. If not designed and deployed with clear ethical guidelines, Generative AI can have unintended consequences and potentially cause real harm. The major risks and dangers of Generative AI can be broadly grouped under technical risks, ethical risks, and societal risks, as discussed in Figure 2-2.

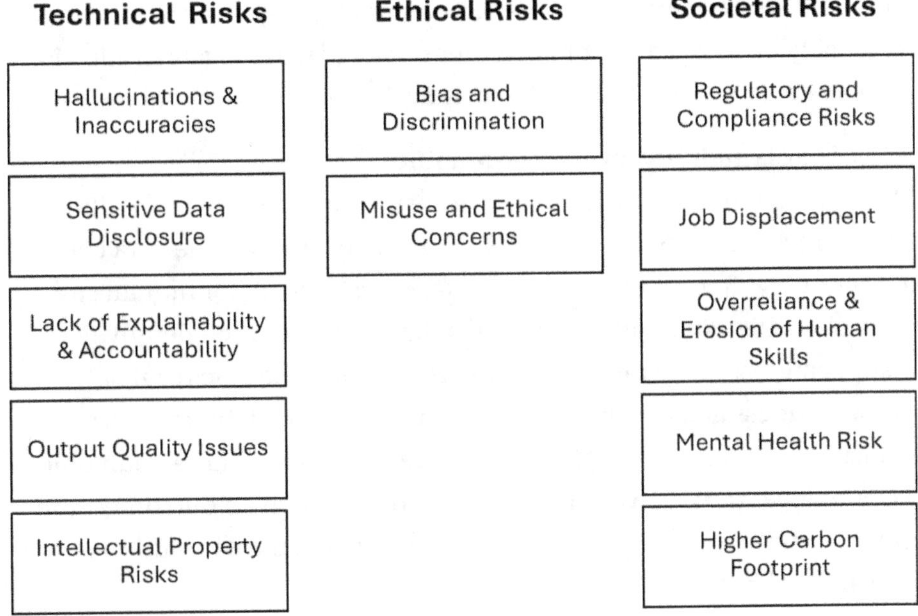

Figure 2-2. *Risks of Generative AI*

Technical Risks

Hallucinations and Inaccuracies

Generative AI models can sometimes create false facts or invent non-existent sources. This phenomenon is known as AI hallucination. Hallucinations can be either harmless or have major legal consequences. Incorrect Generative AI outputs can negatively impact business outcomes, leading to possible legal liabilities. It can also harm the organization's brand image and damage the reputation. To prevent these risks, strong safeguards are needed to ensure Generative AI outputs are accurate and reliable.

Sensitive Data Disclosure

If the training data for Generative AI models included personal information and that is used in the generated output, it can accidentally reveal sensitive personal information. This becomes a privacy issue and also a security concern. Unintended disclosure of sensitive patient data or product strategy information to unintended third-party recipients can irrevocably breach patient or customer trust and carry legal ramifications. This can be dangerous and lead to privacy issues and potential misuse.

Lack of Explainability and Accountability

Generative AI models are trained on a vast amount of data from different data sources, which is not generally known to everyone. The data used for training is often a black box. Therefore, it is difficult for humans to find a plausible explanation for the results produced by Generative AI applications. If we cannot interpret the rationale for the results, trustworthiness goes down. It becomes a bigger issue if there is an adverse impact on human lives and livelihood due to the unexplained outcome of Generative AI. Who would be accountable for such an outcome—the firm that built the AI solution or the end user using the AI solution or the software itself?

Output Quality Issues

The quality of output from different Generative AI models may vary.
While the output from one model may align to the marketing guidelines
of the organization or be suitable for one cultural context, the output from
another may not. A human review is therefore always needed to assess the
output quality.

Intellectual Property Risks

Generative AI models are trained on large amounts of existing data which
can include copyrighted content. It can also learn and further improve
from user inputs during interactions. When such content is produced as
output from Generative AI applications, who owns the copyright for that?
Generating copyrighted content without the permission of the creator can
also potentially lead to legal issues.

The lack of transparency regarding training data in Generative AI
models can cause regulatory implications. Ambiguities over the authorship
and ownership of AI-generated content can raise possible allegations of
plagiarism or risk copyright lawsuits. Any copyrighted material used for
training or fine-tuning the AI models should be clearly disclosed.

Ethical Risks

Bias and Discrimination

Since Generative AI uses LLM models, most of which are proprietary, it
is difficult to assess the quality of data used to train them and check for
any ethical issues. If the dataset used for training the underlying model is
biased, it would lead to biased output. The risk of biased output is higher
when AI is making decisions directly for end users in areas like hiring,
lending, or healthcare.

Training models with diverse datasets, fine-tuning them for fairness,
and implementing bias detection with human oversight and feedback loop
can significantly reduce risks due to bias.

Misuse and Ethical Concerns

Generative AI has massive implications on humans and society. It is affecting our everyday lives with a significant impact on our well-being and how we perceive the outside world. Misuse of Generative AI for various purposes raises many ethical concerns. Below are some questions around the ethical use of Generative AI:

- Should Generative AI be used by students to complete their assignments that they are supposed to do?

- Should it be used to create deepfakes or clone voices without consent?

- Who owns the copyright to AI-generated content?

Using Generative AI to create fake evidence, spread malware, or post false, misleading, and biased content on social media or generate toxic content to harass individuals online raises questions around its ethical use. Malicious use of Generative AI can have a detrimental impact on society. Hence, robust ethical principles are necessary to guide AI development and implementation.

Societal Risks

Regulatory and Compliance Risks

The technology advancements for Generative AI are happening at an unprecedented speed. However, the government and regulatory bodies are still catching up with the advancements. This brings in risks if the applications built using Generative AI do not comply with the slowly evolving laws. Complying with data privacy regulations like GDPR, HIPAA, DPDP, and other privacy laws can be even more challenging if Personally Identifiable Information (PII) data is not handled carefully.

Job Displacement

Generative AI has the potential to displace or replace human creators at a pace that was never expected. Human beings involved with ethical decision-making processes may also be at risk of being replaced due to AI-enabled automation. All of this can negatively impact the morale of the workers. The workforce should be trained on new skills and equipped for new roles required to adapt to this change. Otherwise, it can lead to job losses.

Overreliance and Erosion of Human Skills

Dependence and excessive reliance on Generative AI can reduce critical thinking, creativity, and verification skills of human beings. Excessive trust can lead to human oversight and poor decision-making skills. Using AI at every step to do all activities will have a negative impact on the learning and development of core skills. Using AI tools as a shortcut can hamper creative thinking and problem-solving abilities. People may forget how to use formulas for problem solving. Students using AI for language translation may not properly develop their language understanding. They may not realize this in the short term, but it does have a potential long-term negative impact. All should be educated on how AI could hinder mental development and encouraged to embrace it ethically and responsibly within the set of guidelines.

Mental Health Risk

According to research published by the American Psychological Association, excessive use of AI technologies can impact users' mental health. Users interacting more frequently with AI systems may more likely experience loneliness, insomnia, and increased after-work alcohol consumption.

Higher Carbon Footprint

Developing and training large AI models requires significant computational power, leading to high energy consumption and environmental concerns. The energy consumption of AI model training is of growing concern. Sustainable AI practices are needed to reduce the environmental impact of AI computing. Energy-efficient hardware and optimized training processes can reduce the impact.

As per MIT research, LLMs built using efficient neural network architecture can reduce carbon footprints. Using data centers powered by renewable energy sources to train the models and execute AI workloads can be another approach to minimize carbon footprints. According to the CEO of OpenAI, they are spending tens of millions of dollars on energy costs just to process "Please" and "Thank you" in the user prompts.

Misuse of Generative AI

The incredible versatility of the foundation models is also what makes it difficult to predict exactly what kinds of undesirable output they might produce. The capabilities of Generative AI can potentially introduce new risks if misused. It has raised many new questions and concerns of different magnitudes and urgency. Questions have arisen around ethical issues and risks around data privacy, security, misinformation, plagiarism, copyright infringements, harmful content, to name a few. Figure 2-3 shows some of the most common misuses of Generative AI.

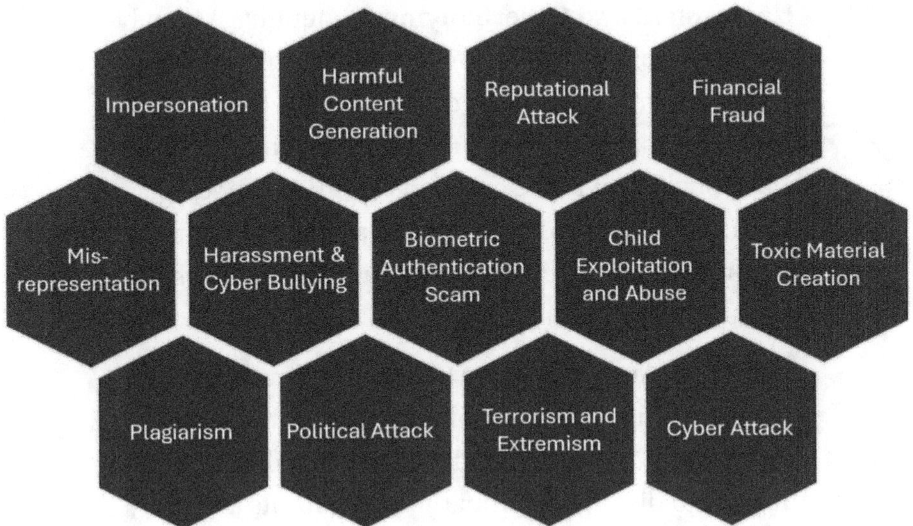

Figure 2-3. *Possible Misuses of Generative AI*

- **Impersonation**: Generative AI can mimic the unique writing skills and linguistic patterns of an author, the voice or tone of a singer or replicate styles or trends of popular designers or another successful product or a retro design.

- **Harmful Content Generation**: Content with offensive language and incorrect information can be generated by AI to misguide people.

- **Reputational Attacks**: Audio or video created by Generative AI can impersonate individuals, manipulate public opinion, and conduct sophisticated social engineering attacks to defame or disgrace others.

- **Financial Fraud**: Generative AI can be used to create deepfake videos, voice, or content which can bypass security measures and can be used for identity theft or trick people into doing financial transactions like wire transfer.

- **Harassment and Cyberbullying**: Generative AI can be used to automatically create harassing or threatening messages, emails, or posts on a wide variety of platforms that can cause substantial psychological and emotional harm to the targeted victim. AI-powered autonomous troll bots can greatly amplify the impact of harassment and cyberbullying and create a hostile online environment.

- **Biometric Authentication Scams**: Generative AI can create highly sophisticated fraudulent materials like synthetic identities and fake documents that can be hard to distinguish from the originals. Its ability to generate such identities and documents at scale increases the magnitude of fraudulent attempts that institutions have to handle.

- **Child Exploitation and Abuse**: Generative AI can be used in a way that can sexually exploit and harm children. Perpetrators can use AI enabled apps easily to create fake imagery, including synthetic media, digital forgery, and nude images of children and upload them to the dark web.

- **Plagiarism**: Generative AI can reuse ideas and steal core concepts from copyrighted work and produce new content without giving credit to the original creator and thus may be infringing copyright and trademarks. Whether the use of Generative AI would be considered as cheating would depend on the task performed and the intent to use the technology.

- **Political Attack**: Misleading political ads and content generated by AI can amplify ongoing election misinformation issues. Bad actors can use Generative AI to endanger democracy or incite violence by using it to produce false, inaccurate, misleading, and biased content. AI-generated deepfakes can pollute the elections too.

- **Terrorism and Extremism**: Terror groups can use Generative AI to rapidly create extremist content and spread propaganda to promote their values and beliefs through social media platforms, websites, and messaging apps. They can generate fake images, videos, or audio clips with hate speeches and radical ideologies which can increase the breadth of propaganda produced, intensify messages, and affect people's attitudes and emotions.

- **Cyberattack**: Cyber criminals can produce more convincing phishing emails and deepfake technologies to trick victims using Generative AI. It can also be used to create advanced malware that can be hard to detect and defend using traditional defense mechanisms.

- **Misrepresentation**: Generative AI tools can be used to easily and quickly create massive amounts of false information and fake content. It can be used to imitate the voices of real people and create pictures and videos that can be hard to distinguish from real ones.

- **Toxic Material Creation**: Actors with bad intentions can use Generative AI capabilities to design chemical formulas for toxic compounds that can be harmful for mankind.

Real-Life Examples of Generative AI Misuse

Generative AI can be used to create your digital twin that looks and talks like you, making it difficult to differentiate from the real one. These digital twins, when used with any malicious intent, can pose several challenges that will be hard to fend off for customers, businesses, government organizations, and many more. Generative AI has made many tasks simpler and more accessible than they used to be in the past. This has opened new concerns, especially around security and ethics.

Generative AI was used by fraudsters to impersonate co-workers on a digital video call. This caused a worker in a financial company to fall prey to the fraudsters, who convinced him to transfer a large sum of money by making him believe that they were genuine.

Generative AI can be used for impersonating public figures, using synthetic digital personas to simulate grassroots support for or against a cause ("astroturfing") and creating falsified media.

Generative AI Misuse in Finance

Generative AI has made it easier to generate fake content. It has opened endless potential to magnify the scope and nature of fraud in the banking industry, especially with deepfakes. In January 2024, an employee of a Hong Kong-based firm was tricked by fraudsters to send US$25 million to them. The fraudsters created a deepfake of the Chief Financial Officer (CFO) that instructed the employee to execute the financial transaction on a video call in the presence of other colleagues, which also turned out to be deepfakes of real colleagues who were never present during the call.

Fraud losses enabled by Generative AI are likely to increase rapidly over the years and is forecasted to touch billions of dollars. Banks and financial institutions are challenged by the pace of innovation to stay ahead of the fraudsters. The new Generative AI tools have made it easy for fraudsters to create deepfake videos, generate similar sounding voices, and fake documents. Large banks like JPMorgan and Mastercard have already started investing significantly to build AI/ML-based solutions that can detect and build solutions against frauds.

Generative AI Misuse in Government and Politics

Generative AI can be misused to produce misinformation and disinformation at scale. It can be misused to create convincing but false narratives, press releases, and government announcements. AI-generated fake audios/videos can impersonate public officials leading to panic and confusion. It can also influence voter sentiments with misleading political content or fake endorsements. When AI-generated misinformation overwhelms real information, it's hard for the public to trust what's true. This can corrode social and political trust.

People can misuse Generative AI to create fake government IDs, permits, or compliance certificates to breach the government systems. LLMs trained on sensitive PII can inadvertently expose private information. Generative AI, if trained on flawed data, can amplify societal bias, leading to discriminatory welfare decisions or biased law enforcement. It can be used to send personalized phishing emails or develop malicious code to compromise and attack sensitive government systems and public infrastructure.

Generative AI Misuse in Education

In the educational sector, plagiarizing others' work or passing off AI-generated content as their own is one of the biggest concerns. It is also a violation of IP copyrights. Any content that is generated using AI tools should be called out to ensure that work is not incorrectly credited to someone. Students may use AI to bypass plagiarism software by using AI to rewrite others original contents. They may use AI to solve complex mathematical problems or complete their assignments or even generate the code for their programming assignments.

Generative AI Misuse in Media and Entertainment

Misuse of Generative AI in the media and entertainment industry can spread misinformation, cause ethical violations and damage to reputations. Below are some examples:

- **Deepfake Videos and Images**: There are instances where deepfake videos of Tom Hanks and MrBeast were used in scam advertisements promoting fake products. This misled consumers and caused damage to celebrity reputations. Also, a deepfake video of the former president of the United States, Barack Obama showed him saying things that he never actually said. This spread false information

- **Fake News and Misinformation**: A deepfake video of Ukrainian President Volodymyr Zelenskyy falsely announcing surrender to Russia went viral. This misled the public. But such misinformation and fake news can manipulate public opinion and cause social unrest.

- **Copyright Infringement**: AI tools have been used to generate content that infringes on copyrights, such as creating music, art, or literature that closely mimics the style of existing artists. AI-generated tracks that imitate the style of a famous musician could potentially devalue the original artist's work.

- **AI Voice Cloning for Malicious Intent**: Scammers used AI-generated voices of famous musicians like Drake and The Weekend to create fake songs that went viral. This misled fans and deprived the original artist of their royalties.

- **AI-Generated Fake Actors in Films and Ads**: Some companies have used Generative AI to create fake actors without disclosing that they are not real people. In China, AI-generated news anchors have been used to present news, blurring the line between reality and AI. This raises ethical concerns about transparency.

Generative AI Misuse in Healthcare

Generative AI can be misused to generate entirely false news articles about medical breakthroughs, cures, or dangerous side effects. These articles may sound convincing, potentially causing panic, spreading misinformation, and eroding trust in legitimate sources. Additionally, Generative AI can create deepfake videos of renowned doctors or healthcare providers giving false medical advice or endorsing unproven or harmful products.

For example, an AI-bot of Facebook claimed that Chia seeds can help get diabetes under control. The video received over 40,000 likes, was shared more than 18,000 times, and generated over 2.1 million clicks. Though chia seeds have medicinal values, there is no scientific evidence that chia seeds could cure diabetes or help get it completely under control.

Generative AI Misuse in e-Commerce and Retail

While Generative AI has introduced many use cases in the retail industry that are helpful to mankind, it has also introduced many loopholes and challenges that e-commerce and retail players are struggling with. Below are some examples of misuse of AI in the e-commerce and retail industry:

- **Fake Reviews and Listing**: Amazon, for example, had challenges on how to deal with AI-generated fake product listings and reviews. Such counterfeit listings damage brand reputation and cause loss of sales for authentic sellers. Fake reviews mislead the customer about product quality and cause customer dissatisfaction.

- **Deepfake Advertising**: In 2023, AI-generated deepfake videos of celebrities like Tom Hanks were used to promote scam products without their consent. It misled customers and violated intellectual property rights.

- **Scam Ads and Phishing**: Fraudulent sellers have used AI to create professional-looking but fake ads, tricking users into purchasing non-existent products. Scammers are using Generative AI to create fake emails and chatbot clones of reputed brands like PayPal, tricking customers into sharing payment or personal information. Google and Meta struggled to remove AI-generated scam ads promoting fake retail sites.

Generative AI Misuse in Cybersecurity

Generative AI can be used by hackers to write malicious code, such as ransomware, keyloggers, or trojans. It can also be used to engineer social engineering attacks with phishing emails or messages that perfectly mimic legitimate senders. It can be used to trick employees into clicking malicious links, downloading malware, or revealing credentials. It can be used to create deepfakes with cloned voices or videos that impersonate senior executives and authorize wire transfer or data access.

Goals for Misuse of Generative AI

Actors have specific reasons or motivations in mind to misuse or abuse Generative AI capabilities. It can range from financial gains and scams doing fraudulent activities to harassment and political disruptions. Understanding these motivations is important to assess the downstream impact and to plan for appropriate countermeasures. According to research done by DeepMind, the following are some of the main reasons for misuse of Generative AI capabilities:

- **Deception and Misinformation**

  - **Influence public opinion** to distort public's general perception of sociopolitical realities by impersonating public figures

  - **Defamation** to portray political figures or dissidents in compromising situations that can spoil their reputation

    For example, in 2022, an AI-generated deepfake falsely showed Ukraine's President Zelenskyy urging soldiers to surrender during the Russian invasion before it was debunked.

- **Financial Gains**

  - **Monetize products and services for profit** using Generative AI to create large volumes of low-quality content like books, articles, and product ads for sale on online platforms to maximize business profitability

  - **Scam and fraudulent activities** to promote fraudulent crypto and investment schemes using AI-generated images and videos of influential figures

- **Impersonation and Identity Theft**

  - **Impersonate trusted individuals** like loved ones or senior executives using AI-generated audio or video to steal sensitive information or extort money from victims

    For example, scammers used Generative AI–based voice cloning to impersonate a senior executive and instructed a finance employee to transfer funds urgently for a "confidential acquisition."

- **Defamation and Reputation Damage**

  - **Harass or intimidate** individuals by creating and sharing AI-generated intimate imagery of adult or adolescent females without their consent as part of a bullying campaign or in an attempt to defame or for extortion.

    For example, a Tom Hanks deepfake scam led to a 10% drop in fan trust, per a 2024 survey.

- **Propaganda and Ideological**

  - **Promote terrorism and extreme ideologies** to create civil unrest and political violence with AI-generated hate speech and extreme narratives

    For example, in 2023, an AI-generated image of an explosion outside the US Pentagon went viral, briefly causing confusion and a market reaction before authorities confirmed it was fake.

- **Cybercrime**

  - **Conduct cyberattacks** to disrupt technical systems and networks using AI-generated malware

  - **Identify high-value organizational targets** and their vulnerabilities using AI-generated code

Characteristics of Generative AI That Can Be Misused

Table 2-1 summarizes the different characteristics of Generative AI and how they can possibly be misused.

Table 2-1. *Generative AI Characteristics That Are Misused*

Generative AI characteristics	How it can be misused
Human-like content generation	• Generate disinformation and fake news • Create deepfakes for impersonation • Create realistic phishing emails, fake websites
Speed and scalability	• Generate automated spams and phishing emails • Flood social media or forums with automated posts to create a false sense of consensus • Generate and send harassing messages using Bots
Mimicry abilities	• Impersonate by cloning someone's writing style or voice • Create counterfeit content • Plagiarism and copyright infringement
Personalization capabilities	• Hyper targeted manipulation • Social engineering scams • Exploit known vulnerabilities

Common Strategy for Misusing Generative AI

Generative AI can be misused in broadly two ways—misuse the capabilities of Generative AI or compromise the Generative AI system. The former is the easier and most common way to put Generative AI into wrongful use like creating deepfakes or phishing emails, as it doesn't need advanced technical skills. Using Generative AI for impersonation, scams, and synthetic personas has now become easier due to easy access to Generative AI tools at a reduced cost. For example, scammers used AI to impersonate a CFO, costing a Hong Kong firm $25M in 2024. This has led to increased concerns over use of Generative AI for disinformation, image cultivation, defamation, and other fraudulent activities that are detrimental to society.

If you look at history, malicious actors have used digital technologies for defamation, scams, disinformation, counterfeit or falsification, non-consensual sexual abuse, or child abuse. Now with easy access to Generative AI tools and technologies, such misuse for harmful intent can possibly be on the rise. If Generative AI capabilities are misused to depict a real person in real time, it carries a higher potential to harm with disastrous impact. For example, using Generative AI to create a deepfake video depicting a real politician giving a hate speech can potentially cause riots or trigger communal violence in society. Similarly, a static image created using Generative AI can be made viral to portray incorrect information. Using Generative AI to create non-consensual intimate imagery or child sexual abuse material can lead to defamation of individuals.

Creating realistic images or videos of non-human objects like documents, songs, events can potentially lead to IP infringement if Gen AI is used to produce parts or the entirety of someone's intellectual property—such as literary and artistic works—without permission. Generative AI can be misused to create counterfeit replicas of original art or music composition and falsely claim them as authentic. It can be used to fabricate objects, events, or places and claim them as real.

Ethical Development Guidelines for Generative AI

To reap the true benefits of Generative AI, organizations must have a clear and actionable framework for governing the responsible use of Generative AI. The growing momentum and the rapid mainstream adoption is driving the need for responsible AI—developing, deploying, and using AI in an ethical manner. Guidelines for AI ethics must focus on the following core UNESCO values for the goodness of humanity, individuals, societies, and the environment throughout the life cycle of AI systems:

- **Human Rights and Human Dignity**: Respect, protect, and promote human rights, fundamental freedoms, and human dignity.

- **Peaceful and Just Societies**: Unwanted harm and vulnerabilities to attack should be avoided.

- **Diversity and Inclusiveness**: Respect, protection, and promotion of diversity and inclusiveness should be ensured.

- **Environmental Flourishing**: Environment friendliness, ecosystem protection and restoration, and sustainable development should be promoted.

Principles for Responsible and Trustworthy Generative AI

All Generative AI applications should be governed by the following principles throughout the life cycle. Figure 2-4 illustrates the key principles of building responsible and trustworthy Generative AI systems.

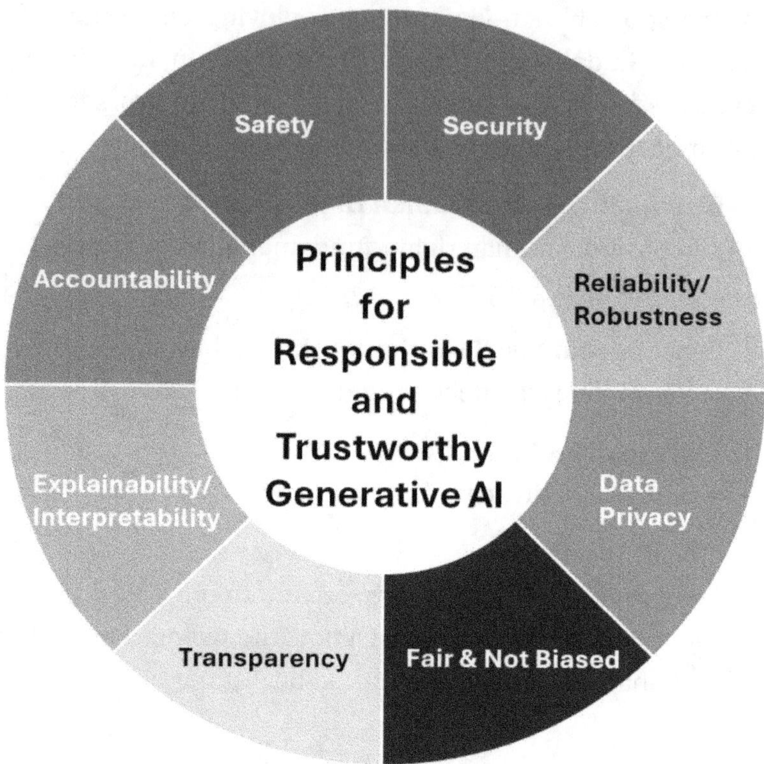

Figure 2-4. *Principles of Responsible AI*

Safety

Generative AI applications should at all times ensure the safety of human life, property, and environment. Safety of the Generative AI application should be enforced through responsible design, development, and deployment practices. Deployers and end users of the application must use the application responsibly. End users should be made aware of the risks, if any, through adequate and clear explanations and documentation. Safety considerations must be employed throughout the life cycle, starting as early as possible from planning and design to prevent failures or conditions that can render the system dangerous. Organizations must conduct a thorough risk assessment and develop a contingency plan to address any potential risks in the AI application.

Security

Generative AI systems should be protected from various cybersecurity vulnerabilities like tampering with training data or data theft through AI system endpoints. Confidentiality, integrity, and availability must be maintained through protection mechanisms that prevent unauthorized access and use. Rigorous testing and model validation along with cybersecurity measures are essential to mitigate security risks. Alongside security, Generative AI applications must be robust and resilient enough to withstand any unexpected changes in the ecosystem or adverse events. Continuous monitoring and security updates to safeguard against emerging threats and vulnerabilities are essential for enhanced security and robustness of the systems. For example, JPMorgan uses AI to monitor network traffic and detect tampering attempts in real time.

Data Privacy

Generative AI models are trained on large sets of data. They also take input data as prompts. Protecting sensitive user information in the training data or user-input data is important to ensure privacy, a right essential to protect human identity and dignity. All data must be collected, used, shared, archived, and deleted as per the regulations to prevent breaches and maintain trust. For example, Google anonymizes user data in its AI ad tools to comply with privacy laws.

A robust data protection framework and governance should be implemented to ensure data security and privacy. Data minimization, anonymization, and encryption with secure data storage techniques ensure customer data stays secure, building trust and avoiding legal penalties. Explicit consent should be obtained for the collection and processing of sensitive personal data. Clear policies for data retention should be established to prevent prolonged storage of unwanted information. The EU AI Act, effective 2025, mandates strict data protection for AI systems.

Reliability and Robustness

Robustness and reliability of AI systems is essential for its widespread adoption. Availability and consistency are paramount for a trustworthy AI system. The output of AI systems must be consistent and reliable even under high volumes of usage. Even when it fails, there must be some degree of predictability and consistency. Regular testing with diverse datasets ensures AI outputs remain stable. Imagine using AI to analyze brain scans for medical diagnosis. If the analysis and output are inconsistent, it can risk human lives.

Regular monitoring is needed to ensure robustness and reliability. It must ensure that AI algorithms produce the right results for every new dataset. There should also be established processes to handle issues and inconsistent results when they arise.

Fair and Not Biased

The challenge with fairness and being unbiased is even more with AI as it lacks nuanced understanding of social standards. A biased output can be technically correct but socially unacceptable. Hence, the outcomes of Generative AI should not be influenced by bias based on gender, age, race, color, language, religion, location, or any other societal and cultural factors. Bias can normally come from datasets used for training the models and can lead to unfair outcomes. Even biased models can generate output that is unfair or discriminatory toward certain groups. This can lead to erosion of trust in AI systems.

Using a curated and diverse dataset augmented with bias-detection algorithms can ensure fairness in AI-generated content. Algorithms should be checked and re-weighted to remove biases in them. Ensuring diversity in teams and establishing ethical practices for development and deployment of AI systems is crucial to identify and address bias. Continuous monitoring and evaluation of AI systems can also identify biases, if any.

Transparency

Accountability and Transparency form the foundations of trustworthy Generative AI systems. Stakeholders should be informed of how the AI systems work, make decisions, or generate outputs. They should be informed of the expected quality of output and its potential impact. Promoting higher levels of transparency increases trust and confidence. The need for transparency increases when the outputs are incorrect or could have a detrimental impact on human life. How can you explain Google Maps showing direction over a bridge that is still under construction?

The scope of transparency covers the entire life cycle of Generative AI—starting from design decisions and training data to how and when the application is developed, deployed, and used by end users. Transparency is not just about providing visibility to source code but should provide information about the algorithms and data used to train the models making the decisions. It should also inform about the broader socio-technical impact of using AI-generated output. Warnings around unexpected behavior of systems should be clearly highlighted. Proper documentation should be provided for the data used, the model's behavior in different scenarios, and the potential biases impacting the outcome. Any known limitations and biases should be documented. End users should always be informed when they are interacting with Generative AI-powered systems. Providing users with the ability to control or provide feedback for improvements is another way to promote transparency.

Explainability and Interpretability

Explainability and Interpretability can shed light on the output of Generative AI models. The results should be explainable and understandable for humans. Explainability refers to the ability to describe in simple terms why and how the AI model generated the output. The documentation should provide details about the models used for the implementation and the source of data used for training. Interpretability, on the other hand, refers to how easily a human can understand why a machine learning model made a decision. Explainability doesn't require end users to understand how the model works. But users must know why an AI-generated a specific output from the documentation of the models and data sources.

Appropriate levels of information about the rationale behind the output of generation AI systems should always be available to individuals interacting with it. Without that, the users cannot verify and evaluate the AI-generated output.

Accountability

Trustworthy AI solutions must include policies and guidelines that specify who is responsible and accountable for the output of AI. This is important when AI is used for medical diagnosis or for autonomous car driving. Who should be held responsible if AI output leads to misdiagnosis impacting the life of a human being? If an AI-powered self-driving car causes a collision, who is responsible and accountable for the damage - the driver or the vehicle owner or the manufacturer or the AI programmer or the CEO? Similarly, if AI-powered trading platforms lead to losses, who should be held accountable?

It is important to identify the laws and regulations that will determine the legal liabilities in such cases. The questions that need to be answered are should AI systems be covered under existing laws, how should the negative impacts of AI be communicated to public and law enforcement authorities, and what would be the consequences for the responsible parties.

Approach to Prevent Misuse and Mitigate Risks in Generative AI

Potential for Generative AI across industries is huge. But when in the wrong hands, it can be disastrous. So how can we prevent potential misuses? The answer is to build a strong governance for Generative AI that not only prevents potential misuses but also promotes its responsible usage to balance reward over risks. The governance principles should be applied at different stages of the life cycle starting from strategy, planning and design through data collection, algorithm selection, and model training to deployment and rollout.

Below we look at how to apply the different principles throughout the life cycle of Generative AI application development to build trustworthy and responsible solutions.

Risk-Based Use Case Prioritization

Use cases for Generative AI implementation should be identified and prioritized following a risk-based scoring approach. The associated risk and impact of the use case must be assessed. Use cases with low risks should be prioritized. Use cases should be humane and have clear societal and business values. Ensure that they cannot be misused or used to promote bias or misinformation. Avoid or heavily restrict high-risk use cases unless robust safeguards are in place. Innovate responsibly by focusing on low-risk and high-value applications.

Establish Guardrails

Policies must be defined that will ensure that Generative AI does not generate any harmful content at any point of time. Nobody should be able to abuse the AI algorithms to generate unfiltered responses or harmful advice or recommendations. The use case for Generative AI should not generate any unacceptable output. For example, a Generative AI product used for children should have a policy that prohibits generation of violent or adult content. However, the defined policies should also consider handling exceptional situations. A well-defined policy with proper guardrails is fundamental for building responsible AI applications.

Control Access and Safeguard Content

Implement a tiered, role-based access control for Generative AI tools and technologies and applications built using Generative AI. Full access should be given only to trusted users, while others get restricted or limited access. Monitor output to detect misuse patterns early. Track who is using AI for what purpose. Suspicious usage patterns must be flagged.

Watermarking should be used to identify and track the provenance of AI-generated content. Bulk generation of hate speech, misinformation, or harmful/toxic content must be prevented.

Bias Detection and Mitigation

Foundation models and fine-tuned models must be regularly tested for bias against gender, age, race, nationality, etc. Training datasets should be diverse and inclusive enough to ensure unbiased output. Test the application for various possible misuse scenarios through simulations. All critical decisions and outputs involving legal, health/medicine, and finance must include human oversight.

Strong Governance and Regulations

Organizations must have a strong governance for their Generative AI initiatives. Organizations must have a cross-functional AI ethics board that regularly reviews and approves AI releases. The AI governance committee should perform regular ethical impact assessments for new use cases.

The governance framework must define the approach to design, build, test, and deploy responsible and ethical AI solutions. Generative AI practices must prioritize transparency, fairness, and accountability. It must ensure compliance with regulations and laws specific to the industry and geography. All adverse events or incidents caused by Generative AI must be reported and handled as per defined escalation protocol.

Safety and Security Policies

Rigorous safety protocols must be followed while implementing Generative AI solutions to prevent any harmful or unintended consequences. Strong data protection measures should be implemented to respect and uphold user privacy. Any sensitive or confidential data used for training the models should not be included in the output to prevent breaches of confidentiality and trust. Human oversight is essential to monitor the outcomes of AI, prevent human rights violations, and minimize risks to society. Additionally, there should be opportunities for humans to provide feedback and control the quality of AI outputs.

Quality Reassessment

Unlike traditional applications, testing Generative AI applications is very different due to the variable nature of their output. The output for the same prompt may be semantically correct but may not be exactly the same. At times the output may also be incorrect due to hallucination. Generative AI can produce text with falsely attributed quotes, invented data, and supposed "findings" that sound plausible but are not real. This requires a very different approach to test Generative AI applications. Quality assurance engineers must not only follow robust validation processes but also have a keen eye to validate facts to suspect and identify misinformation and remove it from AI-generated content. Quality assurance approaches should include testing for ethical concerns like bias, fairness, and privacy. Evaluating the reliability and trustworthiness of the Generative AI must also be prioritized. Security testing is critical to the success of Generative AI.

Public Awareness and Digital Literacy

Increasing the general awareness about Generative AI tools and technologies within employees should be the first step for any organization. Ban on using these tools to prevent misuse can be rather counterproductive. Organization policy must provide ground rules and practical guidelines for ethical use of Generative AI. Awareness and training sessions highlighting the strengths, limitations, and risks of Generative AI for all employees will help to promote its responsible usage. For example, Microsoft's AI training program educates employees on ethical prompt design.

Conclusion

Generative AI is opening exciting new doors for innovation, efficiency, and growth. At the same time, it introduces some real risks - from misuse and misinformation to ethical and societal concerns that leaders cannot afford to ignore. This technology can boost productivity at scale but without the right controls, it can also amplify harm - Generative AI demands careful governance.

The path forward is balance. Business leaders need to embrace this technology with clear intent and strong guardrails. That means choosing the right use cases, embedding transparency and fairness, and keeping people at the center. Not every problem needs Generative AI - knowing where not to use is as important as knowing where it adds real business value.

Leaders should start with low-risk pilots and build Responsible AI governance into solution from day one. Establish clear accountability and ensure that AI solutions are safe for mankind by establishing an AI ethics board. These steps help ensure AI is used safely, responsibly, and in alignment with organizational values and societal expectations.

With the right leadership and discipline, Generative AI can be a powerful tool for good - driving business growth while making a positive impact on society. The future of Generative AI isn't just about what it can do - it's about how we choose to use it responsibly and accountably. Leadership decisions will play a key role in guiding its usage.

CHAPTER 3

Generative AI in Action: Industry Use Cases

Introduction

Generative AI promises to bring in innovation, enhance productivity, and streamline business operations. Generative AI is being used across multiple industries. It is revolutionizing and disrupting every industry with its versatility and adaptability.

Industry leaders in the manufacturing sector are using Generative AI to refine the design processes for physical objects. Pharmaceutical companies are harnessing Generative AI for the design of proteins and molecules for medical purposes. Hospitality companies are redefining their customer experiences by communicating seamlessly with their customers in various languages in real-time using Generative AI. It is being used for supply chain management to accurately forecast and manage inventory levels. Gartner predicts that by 2026, more than 80% of enterprises would have adopted and deployed Generative AI applications in production for some use case.

B. De, *Generative AI for Business Innovation*, https://doi.org/10.1007/979-8-8688-2682-5_3

This chapter covers the details of how Generative AI adoption is happening in different industries for varied business use cases with real examples from industry leaders.

Generative AI Adoption in Industrial Manufacturing

Generative AI is transforming manufacturing industry by improving efficiency, reducing costs, enhancing innovation, enabling better decisions, and reducing downtimes. Following are some of the common use cases in the manufacturing industry:

- **Optimized Product Design**: Generative AI can be used to quickly create different designs and simulate their performance. It can create 3D designs or digital twins of products that can help manufacturers select the most efficient, optimized, and cost-effective design for their product before manufacturing begins.

  Siemens is using Generative AI to optimize the design of their components. Based on weight-to-strength ratios, material, temperature, pressure, force range, and other parameters, Generative AI algorithms suggest countless design variations to come up with optimal configurations. This cuts the component design time by 40%, saving $10M annually for Siemens, per a 2024 report. Autodesk is using Generative AI to design more efficient and comfortable jetliners for Airbus.

- **Supplier Evaluation**: Generative AI-powered chatbots can interact with potential suppliers asking for price, quotes, and demos. It can evaluate them based on their past data, user reviews, website information, and responses and make recommendations for selection. Generative AI chatbots evaluated 100 suppliers in hours, saving 50% of procurement time per a 2024 report.

- **Production Planning and Inventory Management**: Generative AI can forecast demands by analyzing customer data. The demand forecast can be used to create a production plan based on available materials, equipment, and resources. This can help to mitigate supply chain disruptions and provide recommendations for inventory management and better logistics planning. Overall, this can reduce costs due to overproduction or stockouts and improve overall efficiency. Schneider Electric uses AI to simulate and generate resilient supply chain strategies under various risk conditions, reducing disruptions by 25%, per a 2024 study.

- **Sustainable Manufacturing**: Generative AI can suggest strategies to minimize waste and optimize resource utilization. Generative AI models can forecast the environmental impact of the manufacturing processes and suggest improvements to reduce carbon footprint. Unilever is using Google cloud's Generative AI tools to design eco-friendly product formulations that reduce environmental impact.

- **Improved Quality Control**: Image recognition and analysis can detect defects in products during the manufacturing process, ensuring higher quality and reducing waste. One of the biggest car manufacturing companies, Ford Motors, is using Generative AI to identify wrinkles in car seats.

- **Predictive Maintenance and Assistance**: Unplanned downtime can be costly, leading to revenue losses, production delays, and unsatisfied customers. Generative AI models trained on data from sensors installed on the machines such as temperature, sound, vibrations, pressure, etc. can predict equipment failures. They can continuously scrutinize operational data and forecast potential issues based on observed deviations from normal and recommend preventive maintenance, thus reducing downtime. Honeywell combines Generative AI with IoT sensor data to simulate future equipment states and maintenance needs. According to Deloitte's "AI in Manufacturing Report (2024)," Generative AI can reduce breakdowns by 70% and cut maintenance costs by 25%.

- **Product Documentation and Assistance**: Generative AI can be used to create user manuals, product documentations, and troubleshooting guides. Chatbots trained in product manuals and troubleshooting guides can provide immediate technical assistance to users of machinery or equipment and troubleshoot any issues with them. This can reduce errors and improve productivity. Bosch has used AI-powered service assistants that generate step-by-step troubleshooting instructions in real-time for technicians on the shop floor.

- **Marketing Campaign**: Generative AI can generate tailored and personalized marketing campaigns to promote products. For example, Caterpillar uses Jasper AI, a generative content platform, to craft personalized marketing messages and campaign content for different customer segments.

Generative AI Adoption in Healthcare and Life Sciences

Generative AI is rapidly transforming healthcare industry by supporting clinical decisions and enabling personalized treatments for better patient outcomes, and streamlining healtcare operations.

Figure 3-1 shows some of the common uses in the healthcare domain.

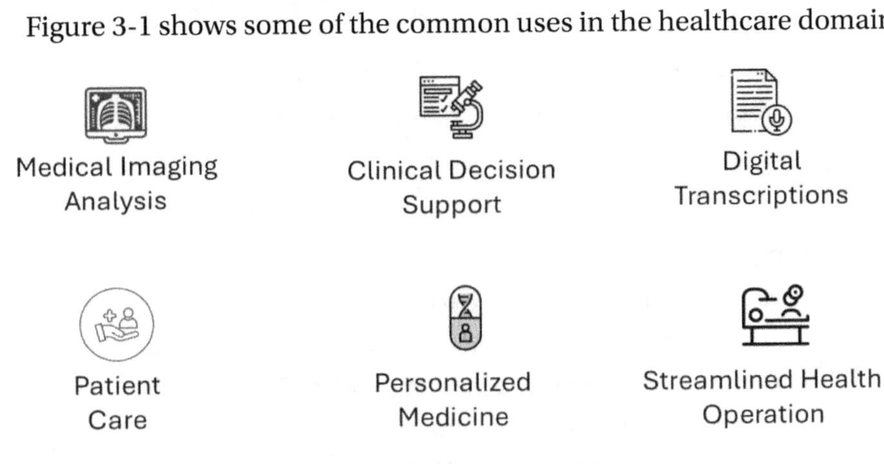

Figure 3-1. *Generative AI Use Cases in Healthcare*

- **Medical Imaging Analysis**: Generative AI can be used to enhance, reconstruct, or analyze images for X-Rays, MRI, or CT scans for medical diagnosis. It can identify early-stage conditions such as brain tumors, Alzheimer's, and diabetic retinopathy by meticulously examining MRI and retinal scans. There have been reports where AI has detected subtle pathological changes that escaped from human observation. Lunit INSIGHT assists radiologists or clinicians in the interpretation of chest X-ray and can detect 10 abnormal radiologic findings with 97–99% accuracy.

According to a post on Reddit, a user shared that ChatGPT helped interpret complex symptoms and medical reports, suggesting a rare genetic disorder that several doctors had previously missed.

- **Clinical Decision Support**: Generative AI can analyze blood reports and suggest possible diagnosis. This can help to reduce misdiagnosis. Glass Health is an AI clinical decision support platform that generates differential diagnosis from clinical inputs.

- **Patient Care**: Generative AI can improve the patient's experience by automating and streamlining appointment booking, patient communication, and triage. It can also be used to create personalized content for patient education. Patients can also engage in interactive learning experience with Generative AI-powered virtual assistants that can provide answers to their queries. NHS UK uses an AI-powered chatbot named "Florence" to guide patients through medication adherence and checkups. Harvard has developed an AI-powered system to triage symptoms and recommends the next steps. It triaged symptoms of 10,000 patients in 2024 and reduced wait times by 20%.

- **Digital Transcription**: Create transcripts with a summary of key information from clinical patient interaction to automate the documentation process. Hospitals can use AI to generate routine progress notes and discharge summaries. Nuance DAX (Dragon Ambient eXperience) by Microsoft listens to doctor–patient conversations and generates clinical notes automatically.

- **Personalized Medicine**: Generative AI can create personalized treatment plans for a patient based on their genetics, lifestyle, and medical conditions. It can also analyze patient charts and flag discrepancies or gaps in care, improving the overall outcome. Tempus, a health tech company, has used Generative AI to analyze clinical and molecular data (including genomics) to recommend personalized cancer treatments. Companies like CloudMedX and Biofourmis are developing intelligent platforms that manage patient data, support home-based care, and provide predictive insights into the course of treatment.

- **Streamlined Healthcare Operations**: Generative AI can significantly reduce time for billing and administrative tasks by using automated medical coding and ensure regulatory compliance. It can integrate with EHRs to reduce medical errors and ensure that important medical information is not missed. It can optimize appointment scheduling by analyzing patient needs and doctor availability, personalized appointment reminders, and follow-ups.

The life sciences industry is adopting Generative AI technologies to significantly reduce time and cost associated with discovery and clinical trials of new medicines. Figure 3-2 shows some of the common use cases for Generative AI in life sciences industry.

Research Assistance
and Reporting

Drug Discovery
and Design

Protein &
Antibody Design

Patient Matching for
Clinical Enrollment

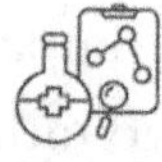

Enhance
Clinical Trials

Pharmacovigilance

Figure 3-2. *Generative AI Use Cases in Life Sciences*

- **Research Assistance and Reporting**: Specialized
  Large Language Models (LLMs) can be used to extract
  information from medical publications, trial data,
  and patient reports efficiently. It can also summarize
  scientific literature and extract valuable information or
  generate documents based on drug discovery research
  data. Pfizer could save up to 16,000 hours of search
  time per year using Generative AI.

- **Drug Discovery and Design**: Generative AI can
  enable the design of antibodies for challenging targets
  such as membrane proteins and discover novel drugs
  to treat rare medical conditions at reduced costs.
  Insilico Medicine used a Generative AI to accelerate
  drug discovery and designed novel small molecules
  for idiopathic pulmonary fibrosis, a relatively rare
  respiratory disease that causes progressive decline in
  lung function. Their AI-designed anti-fibrotic drug
  entered clinical trials in under 18 months, which would

otherwise traditionally take 4–6 years. Even the total cost was reduced to one-tenth of what it would have otherwise cost (around $400 million) had traditional methods been used.

- **Protein and Antibody Design**: Generative models can create novel protein sequences with specific functionalities or properties, which can be useful in protein engineering, enzyme design, and developing novel therapeutics. Google DeepMind released AlphaProteo to design novel proteins that are likely to bind to target molecules.

- **Patient Matching for Clinical Enrollment**: Match patient's health record to inclusion and exclusion criteria for clinical trials using Generative AI to reduce the time to identify suitable groups of patients for clinical trials.

- **Enhance Clinical Trials**: The processes and protocols for clinical trials can be streamlined and expedited using Generative AI. It can analyze data of existing trials and recommend various aspects for future trials such as protocol, location, patient enrollment mode, etc. It can generate different trial strategies and simulate scenarios and compare the results. All of this can shorten the time and reduce the cost for trials significantly.

- **Pharmacovigilance**: Generative AI can simulate molecule interactions and predict possible adverse drug reactions (ADR) which can require hospitalization or even cause mortality. It can analyze genomic and proteomic data and tell how specific genetic groups would respond to the candidate drug. Generative AI can also predict the toxic effects of the drug, if any.

Generative AI Adoption in Banking

Generative AI has the potential to shape entirely new business models in the banking industry. It can be used in banking for different use cases for front-office, middle-office, and back-office operations. With its state-of-the-art capabilities, it can improve customer experience and satisfaction, simplify and automate operations, reduce costs, improve operational efficiency, and manage risks better.

Figure 3-3 shows some of the use cases where Generative AI can be applied in the banking domain.

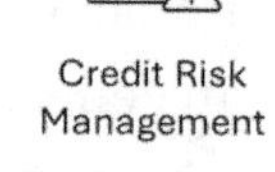

Figure 3-3. Generative AI Use Cases in Banking

- **Personalized Customer Experience**: Generative AI can analyze customer transaction history, spending patterns, and financial objectives and generate bespoke recommendations for investments and services highly tailored to customer needs. This can create more meaningful connections with their customers and drive customer satisfaction and loyalty. Bunq, the second largest neo bank in Europe, has integrated Generative AI in their banking app to allow its users to ask questions about their bank account, spending habits, and savings, with an experience similar to ChatGPT.

- **Enhanced Customer Service**: Generative AI-powered chatbots can interact with customers in their natural language and provide them with the necessary information. These chatbots can provide instant 24/7 support without the customer having to wait in long queues or navigate through complex phone menus. These chatbots can assist customers with a variety of banking tasks, such as getting financial product recommendations, checking account balances and transaction history, completing financial transactions, and more. Wells Fargo has launched a Generative AI powered virtual assistant named "Fargo" to handle everyday banking queries for customers and even perform several other banking operations. It is expected to handle over 100 million transactions annually.

- **Loan and Product Recommendation**: Generative AI can analyze customer's financial data and transaction history and provide insightful data about specific customer behavior and their preferences. This can be used by financial advisors to make hyper-personalized recommendations for loans and financial products. Morgan Stanley has launched a Generative AI-powered assistant that helps its financial advisors to provide hyper-personalized recommendations for investing to its customers based on their account balance, spending patterns, transaction history and financial goals.

- **Personalized Investment Strategies**: Generative AI can generate personalized investment recommendations by analyzing the market trends, financial market data, economic indicators, and investment opportunities. UBS uses Generative AI to tailor and recommend retirement and investment strategies for its affluent customers by analyzing CRM data, life stage information, and preferences.

- **Automated KYC Verification**: Financial institutions can use Generative AI to automate the process for customer KYC assessment by analyzing customer personal data. This can accelerate customer onboarding, reduce false alarms, and enhance accuracy of risk assessment while also ensuring compliance to regulatory requirements for ALM, GDPR, etc. HSBC implemented Generative AI models, to streamline KYC data extraction, validation, and summarization. It helped to reduce review time by 80% for complex corporate on-boarding and improved KYC accuracy and compliance consistency.

- **Fraud Detection and Prevention**: Generative AI can be used to analyze financial transaction data in real time and detect anomalies based on odd location, unusual transaction amount, frequency, and device. It can also be used to generate synthetic data to train fraud detection and anti-money laundering models with diverse datasets and make them more robust and accurate. Alerts can be generated for suspicious activities and sent to banks and customers and help them to take action to prevent them. Mastercard has launched a Generative AI-based model that can help banks detect suspicious or fraudulent transactions with higher accuracy and more quickly.

- **Credit Risk Management**: Generative AI can assess the creditworthiness of a customer more accurately than before based on transaction history, work experience, social data, and other economic factors. This information can be used by financial institutions to make informed lending decisions with greater confidence and reduce default risks. Fintech lenders like Upstart and OakNorth are embedding LLMs into underwriting workflows for faster, compliant credit decisions. Generative AI also can reduce the dependency on manual processes that are often time-consuming and error prone.

- **Regulatory Compliance Reporting**: Generative AI can be used by banks to maintain ongoing compliance with ever-evolving regulatory requirements. Manual and repetitive tasks to interpret new regulations and ensure alignment with regulatory standards can be automated using Generative AI. It can identify compliance risks in their systems and processes and generate timely and accurate reports required to ensure regulatory compliance.

- **Automated Document Processing**: Generative AI is fast changing the back-office document processing by combining techniques like Optical Character Recognition (OCR), Natural Language Processing (NLP), and Intelligent Document Recognition (IDR) for automated document life cycle management. It can be used for document ingestion, classification, and data extraction. It can also summarize contents from contracts like loan policies, regulations, and underwriting. It can also summarize complex regulatory texts to identify and understand the key compliance requirements that organizations must follow. AI can also generate alerts for organizations about new or updated regulations. This helps organizations identify changes that should be implemented to be compliant.

- **Pitchbook Creation**: Pitchbooks are sales materials created by banks to help their employees convince their potential clients to use their products and services. Generative AI can efficiently generate comprehensive and intuitive pitchbooks to drive engaging customer conversations by gathering and summarizing relevant information like economic data and other statistics from different sources.

- **Automated Loan Servicing and Management**: Generative AI is making loan underwriting simpler and faster by automating document verification and risk assessment. This significantly cuts down overall time and efforts for loan processing, thereby improving overall customer satisfaction too. Further Generative AI can be used to generate synthetic data to train models for different types of borrower and risks. This helps banks to build more robust and reliable machine learning models, reducing the overall risks in the underwriting process.

- **Dynamic Contract Generation**: Generative AI can create reviews and update contracts in a streamlined manner. It can produce investment summaries, portfolio insights, or budgeting tips based on customer profiles. It can help bank analysts quickly create reports by researching and summarizing thousands of economic data or other statistics from around the globe. It can also create customized contract templates based on customer data, service terms, and other details. All of these reduce manual work, minimize errors, and improve efficiency. Goldman Sachs uses LLMs to generate client-ready wealth management reports tailored to financial behavior and goals.

Generative AI Adoption in Insurance

Generative AI has huge potential for growth and transformation to revolutionize the Insurance industry. It provides a platform to fundamentally reimagine the operations, products, and customer experiences in the insurance business. Insurance companies can use Generative AI to enhance customer engagement, improve operational efficiency, and increase revenue. Here are some of the key Generative use cases that insurance companies should focus on to reap the true benefits and be the leader. Figure 3-4 shows some of the most impactful Generative AI use cases in the insurance industry.

Figure 3-4. *Generative AI Use Cases in Insurance*

- **Intelligent Claims Automation**: Claims processing is the "moment of truth" for customers and is also the largest operational cost center for insurance companies. Generative AI can automate the First Notice of Loss (FNOL) and fundamentally transform claims process. Virtual agents can gather claim details and photo evidence through empathetic conversations in natural language. Using vision-to-text models can process images of car accidents and property damage and generate initial repair estimates. It can generate claim summaries for adjusters, reducing the processing time at different phases of the claim process. The results reduced claims processing time from weeks to hours, cost reduction by up to 50%, and improved customer experience.

  Touchless claim processing using Generative AI is the future.

Lemonade, a well-known AI-First insurer, has built chatbots for automated claims processing. Its AI bots "Maya" (customer interface) and "Jim" (claims bot) handle policy purchase, customer interaction, and claims processing. It can process and approve a claim in seconds, improving customer experience, and lowering operational costs. Progressive Insurance uses AI technologies to analyze accident images and estimate vehicle damage, which helped reduce manual assessment efforts and led to faster claim settlement.

- **Underwriting Copilot and Intelligent Risk Assessment**: Underwriters spend significant time processing documents like medical reports, financial statements, and property inspection reports and flag specific risk factors to be considered for underwriting. Generative AI can analyze large volumes of documents, summarize, and assess risk. This also helps to quickly arrive at underwriting decisions. Generative AI can act as a copilot for underwriters and transform underwriting from manual review to AI-assisted risk intelligence, giving a competitive advantage.

  Zurich Insurance uses Generative AI for underwriting and risk insights. It not only simplifies the submission process for brokers but also helps insurers understand risk profiles of clients better.

- **Intelligent Fraud Detection and Investigation**: Insurance fraud continues to rise with increasingly sophisticated techniques. Generative AI can detect "synthetic fraud" by identifying suspicious patterns in claim descriptions or spotting doctored images and invoices. Proactive fraud prevention can reduce fraud losses and ensure stronger regulatory compliance.

Allianz has developed AI systems that analyze claims narratives and detect suspicious patterns. Their "Incognito" AI tool helped detect fraudulent claims and saved about £1.7 million.

- **Hyper-Personalized Customer Experience**: Modern customers expect simple, digital, and hyper-personalized experiences. Insurance companies can use Generative AI to build chatbots that can answer customer queries 24/7 and explain complex policy documents in simple language. It can analyze customer data and make personalized policies recommendations. This increases customer engagement and retention.

- **Enterprise Insurance Knowledge Assistant**: Insurance companies manage huge volumes of documents, including policies, claims files, underwriting guidelines, and regulatory documents, which may be difficult for brokers and agents to process. Generative AI can search across all insurance documents and assist agents, underwriters, and claim adjusters to make decisions faster. It also reduces the training times for new employees.

- **Agent and Broker Productivity Assistance**: Insurance agents and brokers can significantly benefit from AI copilots. AI co-pilots can compare policies, generate sales pitch, summarize customer meetings, and even generate a proposal. This can improve productivity and sales.

Generative AI Adoption in Capital Markets

In the capital markets sector, domain-specific LLMs are very effective in classifying news, preparing and filing regulatory reports, retrieving company-specific information, and creating research reports on companies. Figure 3-5 shows some of the use cases for Generative AI in the capital markets domain.

Research and Analysis

Wealth and Asset Management

Algorithmic Trading

Trade and Transaction Validation

Regulatory and Compliance Reporting

Figure 3-5. *Generative AI Use Cases in Capital Markets*

- **Research and Analysis**: Generative AI can summarize news articles, financial reports, social media sentiments, and generate economic summaries with market analysis relevant to investment needs. It can forecast market trends and generate insights based on key economic parameters and historical market data for investors and traders. Platforms like AlphaSense use Generative AI to provide consolidated insights from millions of documents across earnings, broker research, expert calls, and company documents as well as for financial analysis.

- **Wealth and Asset Management**: Financial advisors of wealth management firms can use Generative AI to provide personalized advice for customers based on research reports, regulatory filings, and their financial status and goals. Generative AI can summarize relevant information from multiple sources, which are mostly unstructured, more quickly and accurately. AI summarized 1M documents for advisors, cutting prep time by 30%, per a 2024 study.

- **Algorithmic Trading**: Generative AI can analyze vast amounts of historical market data, trading patterns, news sentiments, and social media trends. Based on analysis and insights, it can make instant trading decisions (buy or sell) that can maximize returns and minimize risks. Because of adaptability and ability to learn, Generative AI models can continuously update themselves based on the changing market conditions. Bots powered by Generative AI can also place orders based on personalized trading rules. This makes algorithmic trading more accurate and profitable and hence an attractive proposition. Goldman Sachs used Generative AI to analyze 10TB of market data in 2024 to optimize trading strategies. This boosted trading success rates by 18%.

- **Trade and Transaction Validation**: Generative AI can be used to identify anomalies and detect suspicious transactions in exchange and report. This can help to detect and prevent fraudulent transactions. Featurespace has launched the world's first Large Transaction Models (TallierLTM) powered by Generative AI for enhanced fraud detection. Generative AI can also generate suspicious activity reports (SARs) in real-time with lesser human effort.

- **Regulatory Compliance Reporting**: Generative AI can automate and enhance the accuracy of reporting suspicious activities and transactions in a timely manner. It can understand regulatory requirements and ensure timely compliance by automatically flagging datasets with potential risks. Additionally, it can generate necessary compliance reports that are more accurate and reliable with reduced errors. Automated reporting and compliance tasks can significantly reduce the burden of manual efforts and administrative efforts.

Generative AI Adoption in Media and Entertainment

Generative AI in the media and entertainment industry provides innovative and engaging user experience. It has streamlined the production process and made content creation faster and more efficient at a reduced cost. It has brought in more experimentation, innovation, and creativity in the field of content development. Figure 3-6 shows some of the major use cases where Generative AI is used in the media and entertainment industry.

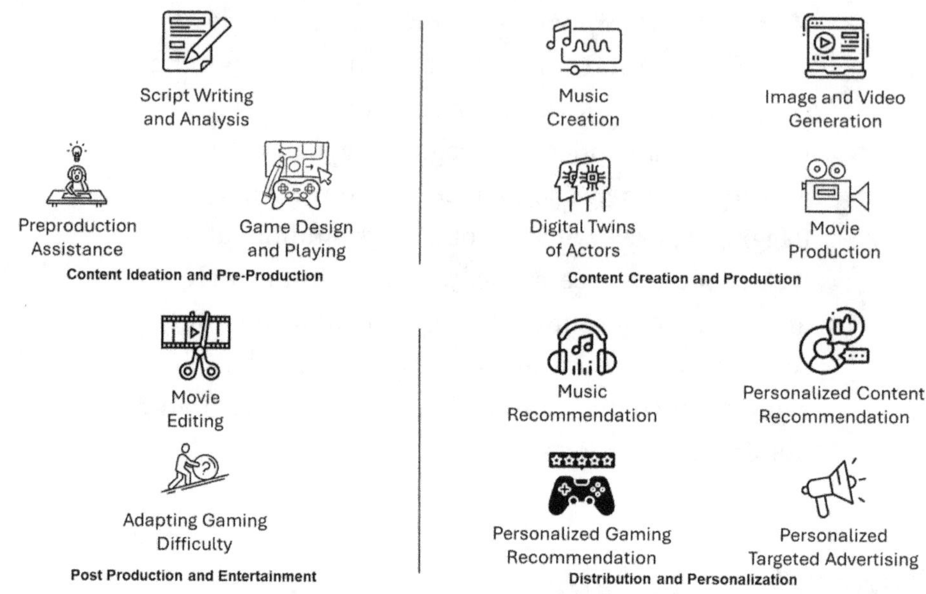

Figure 3-6. *Generative AI Use Cases in Media and Entertainment*

Content Ideation and Pre-Production

- **Script Writing and Analysis**: Filmmakers are using
  Generative AI capabilities to craft impactful movie
  scripts with character backstories with plot twists. By
  feeding the movie's theme and story, Generative AI
  can generate dialogs for all actors in the movie, saving
  valuable time and resources of the film makers. As per
  "Film Production Report, 2024," the screenplay for the
  short film "Sunspring" was written by AI. Generative AI
  can also analyze the script and identify discrepancies or
  questions, if any, helping filmmakers create impactful
  storylines for movies.

- **Preproduction Assistance**: Generative AI can be used to create optimized schedules for the shooting by predicting the task durations. It can suggest suitable shooting locations through virtual explorations. It can also help in shooting by breaking down the script and organizing logical elements to be filmed together or sequentially.

- **Game Design and Playing**: Generative AI algorithms like stable diffusion are being used to create a wide range of themes and virtual items like costumes and gears in games, elevating the overall gaming experience for the players.

Content Creation and Production

- **Music Creation**: Generative AI can learn from existing music patterns to create new and unique music compositions using sophisticated algorithms. AI music generators can create music of different moods and genres and also mix with different instrument styles. However, emotional depth and nuance may spark diverse reactions among audiences. Artists like Taryn Southern have released albums (e.g., "I Am AI") composed and produced with AI. Artificial Intelligence Virtual Artist (AIVA) can compose emotional music for films, games, and commercials. According to a 2024 report, AIVA composed a film score in 48 hours, saving 30% of production time.

- **Image and Video Generation**: Generative AI algorithms like DALL-E, Midjourney, and Stable Diffusion can generate artforms from text prompt, which can be used for creating concept art, storyboards, and marketing materials. It can also transform a given image to other art forms like the ghilbli art. Short video clips and animations too can be created from an input text. Companies like RunwayML offer text-to-video generation capabilities.

- **Digital Twins of Actors**: Generative AI models can be trained on an actor's archive of films and used to create a digital twin or avatars also known as "synthespians," that can replicate the actors' style, voice, inflection, gestures, etc. This can be used to pair the digital twin with the real actor or even de-age the actor as may be required. However, this capability also does raise some ethical questions to ponder upon.

- **Movie Production**: Generative AI can be used to create full-length movies. It can add realistic visual effects, landscapes, and animations. It can create soundtracks, sound effects, and voice-overs. The entire moviemaking process can be streamlined to reduce the time and cost of production. For example, Disney used AI to streamline Avatar 3 production, cutting visual effects costs by 15% in 2024.

Postproduction and Entertainment

- **Movie Editing:** Generative AI can be used to analyze the movie scenes and pick action-packed or emotionally charged scenes to produce exciting trailers for the movie. It can generate custom trailers for movies and TV shows that are tailored to each user's interests. It can also be used for editing full-length films or enhancing the quality of older footage or low-resolution images.

- **Adapting Gaming Difficulty**: Generative AI can be used to analyze the gamers' skills, response time, strategy and dynamically adjust the difficulty level of the game. This makes the game more interesting and less frustrating for the players. Generative AI can balance challenge and accessibility through dynamic difficulty adjustment, providing players with a more personalized experience.

Distribution and Personalization

- **Music Recommendations**: Generative AI can analyze the user's listening history and music preferences and create a music profile for users to make recommendations. The recommendations can be based on genre, mood, artist, and tempo. Spotify uses AI to create personalized playlist based on user's listening history and preferences.

- **Personalized Content Recommendation**: Digital streaming platforms like Netflix, YouTube, and Spotify use sophisticated AI algorithms to generate personalized recommendations for users, based on their history, preferences, and viewing habits. It can analyze a user's viewing history and suggest content that they might be interested in watching.

- **Personalized Gaming Recommendation**: Generative AI can be used to make personalized game recommendations aligned to players' interests based on their preferences, gameplay styles, genre inclinations, and in-game choices. It can also tailor in-game content to provide a more engaging and immersive gaming experience for players.

- **Personalized Targeted Advertising**: Generative AI can create real-time personalized content for advertising. It can predict customer needs by analyzing past data and buying habits and thereby adjust ad placement and timing to make campaigns more effective. Personalized recommendations and tailored content increase customer engagement and conversion rates, making them more effective. Additionally, AI-generated content saves significant time and reduces cost for content production.

Generative AI Adoption in Retail and eCommerce

Generative AI comes with a lot of new potential and offers countless opportunities to enhance customer experience and loyalty. Delivering highly personalized shopping experience has always been a challenge, but Generative AI is proving to be a game changer. Personalization, virtual stylists, catalog automation, demand forecasting, and loyalty programs are some of the main areas that are going to see major transformations. All of this is going to enhance operations and boost sales, driving higher returns on investments.

Figure 3-7 shows some of the use cases where Generative AI can be applied in the field of retail and e-commerce.

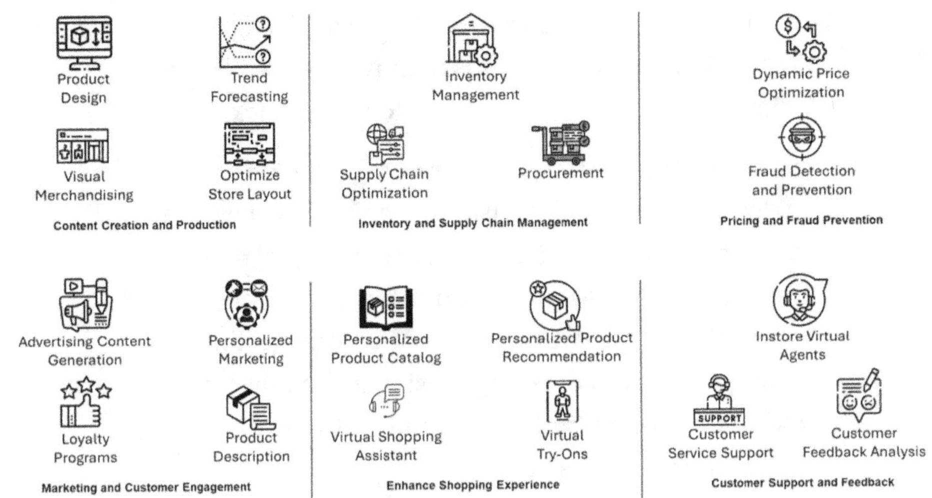

Figure 3-7. *Generative AI Use Cases in Retail and eCommerce*

Product Development and Merchandising

- **Product Design**: Generative AI can analyze current market trends, customer preferences, historical sales data, and recommend new designs for clothes, furniture, or electronics that will attract the customers. It can generate multiple versions of the design that companies can select from. Generative AI accelerates the new design development process helping brands to cater to their customers with fresh and improved designs.

- **Trend Forecasting**: Generative AI can be used to analyze social media platforms, influencer content, fashion blogs, and online customer's shopping behavior to detect emerging trends before they become mainstream. It can also predict future product styles and patterns that are likely to be in demand based on historical data, regional preferences, economic and cultural indicators, as well as climate and seasonal trends. This can help brands manage their inventory efficiently, reducing overstock and understock.

- **Visual Merchandising**: Generative AI can be used to design eye-catching customer-centric visuals for displays that can drive sales. Such displays can be for promotional banners, product highlights, or interactive content. Generative AI can suggest product arrangements or adjust digital signage in real-time within the store based on customers' traffic patterns, demographics and sales data.

- **Optimize Store Layout**: Generative AI can dynamically generate visually attractive and engaging store design and layouts. They can suggest product placements in the store that are likely to grab customer attention.

Inventory and Supply Chain Management

- **Inventory Management**: Generative AI tools can analyze historical sales data, customer behavioral patterns, user sentiments, and market trends and predict inventory requirements to meet future demands. It can produce a detailed analysis of product demand, inventory availability, and emerging trends. This helps businesses to minimize storage costs, prevent stock shortages, and ensure the availability of popular products. AI forecasting reduced storage costs by 20%, saving $2M annually for a retail giant, per a 2024 retail study. Companies like Amazon and Walmart are using Generative AI for demand forecasting and inventory optimization.

- **Supply Chain Optimization**: Generative AI can recommend the best suppliers based on cost, reliability, and delivery performance. It can generate optimized procurement contracts based on pricing trends and supplier behavior. It can also generate optimized warehouse design that can maximize store capacity, organize inventory effectively, and improve operational efficiency. Generative AI can also help to recommend optimal delivery routes that can reduce cost and improve delivery times. AI-optimized delivery routes cut logistics costs by 15% and improved delivery times by 25%, per a 2024 report.

- **Procurement**: Generative AI-powered chatbots can be used to do the initial rounds of supplier negotiations. Procurement associates can use Generative AI to create automated summaries of supplier terms and key insights.

Pricing and Fraud Prevention

- **Dynamic Price Optimization**: Generative AI can do real-time analysis of market data, customer demands, competitor pricing, and customer behavior to suggest a competitive product price that maximizes revenue. It can also enhance B2C customer payments with a customized and dynamic pricing strategy that can enhance the company's competitiveness. It also fosters agility in the constantly evolving market landscape.

- **Fraud Detection and Prevention**: Generative AI capabilities enhance the risk identification and fraud detection capabilities to protect online stores. Traditional AI algorithms can detect and flag suspicious transactions or irregularities in real-time, but Generative AI goes further by simulating new fraud scenarios. It can identify illegitimate purchases and returns or other fraudulent activities. This helps ecommerce businesses to predict and be prepared to prevent such fraudulent activities in the future. It shields themselves and their customers from fraudulent activities.

Marketing and Customer Engagement

- **Advertising Content Generation**: Generative AI tools can design banners, create videos and text for advertising content, and streamline the process for marketers. All of this can reduce the time and money spent on product photography and image editing. Companies like Heinz and Nestle have started using AI to generate content for their product advertisements and promotions. Amazon has been using AI to transform simple pictures of products into more realistic images for customers to see how it would look in different backdrops and environments.

- **Personalized Marketing**: Generative AI can analyze customer's purchasing behavior and preferences and create highly personalized and engaging marketing contents that resonate with every individual. It can ensure that marketing content aligns well with the target audience's preferences. This enhances relevance and increases appeal, thereby maximizing the impact and effectiveness of marketing. It can address email fatigue by suggesting a compelling subject line and creating content that is highly tailored to the individual recipient. Such personalization also helps to strengthen the bond between the brand and the customer.

- **Loyalty Programs**: Retail Industry can revamp their customer loyalty program using Generative AI. They can use Generative AI to analyze customer behavior, their purchasing history, and preferences to create highly tailored reward programs and offers for their customers. They can also create highly personalized email messages, push notifications, and in-app notifications to keep their customers informed about their loyalty benefits.

- **Product Description**: Tools like ChatGPT can create engaging product descriptions that match the brand's tone and style while accurately highlighting the product features, attributes, and specifications. *Phrasee* is another AI-powered platform that helps brands generate, optimize, and personalize marketing content across various digital channels. Shopify has integrated Generative AI in their retail solutions to automate content generation for their products.

Enhance Shopping Experience

- **Personalized Product Catalog**: Generative AI can be used to generate real-time personalized catalog based on their browsing history, purchase behavior, and preferences. It can auto-generate personalized descriptions and images of products that align to customer tastes. It can also adapt catalogs to regional preferences, languages, and cultural nuances. Such a personalized product catalog can help drive customer satisfaction, loyalty, and sales. For example, Macy's used AI to create tailored catalogs, boosting sales by 20% in 2024.

- **Personalized Product Recommendations**: Today's shoppers prefer personalized shopping experiences. By analyzing the data about customer preferences, purchasing behavior, and buying history, Generative AI can provide highly personalized recommendations for products that suit the individual lifestyle and budget preferences. This can raise the chances of sales, improve customer satisfaction, and foster loyalty. It can even recommend ideal discounts that can delight the customers and influence their decisions. Generative AI can also predict a customer's future needs with high accuracy and include it in the recommendations. Amazon uses Generative AI for making exquisite product recommendations to its customers that has accounted for 35% of its product purchases. Stitch Fix, an online personal styling service, uses Generative AI to recommend clothing and accessories tailored to each customer's style and preferences.

- **Virtual Shopping Assistant (Product Search and Discovery)**: Generative AI powered chatbots can interact with customers and answer their queries in natural language while guiding them through their shopping journey. Such a virtual shopping assistant can help to find products and provide a hyper-personalized shopping experience with tailored recommendations and query resolutions. Walmart's AI chatbot assisted 5M customers in 2024, improving satisfaction by 15%.

- **Virtual Try-Ons**: Generative AI can help customers to virtually try products like clothing, makeup, or furniture to see how it fits and looks on them or in their homes before buying. Customers can click their pictures or take videos, and AI can re-generate their 360-degree product views or interactive product demos with the product to create an immersive and memorable shopping experience. This can help brands to differentiate themselves from their competitors and build trust with their customers

Customer Support and Feedback

- **Instore Virtual Agents**: Generative AI can be used to create virtual agents that can engage with customers in multilingual natural language to answer customer queries about products or make recommendations based on their needs. It can also suggest complementary products, create engaging product descriptions, and help customers make informed decisions. Virtual agents can also guide customers and employees in real-time to the right aisles or shelves where they can find their product. Lindex, a Swedish retailer, announced the release of the "Lindex Copilot" to support its store employees with personalized advice for store operations and information about daily tasks. Overall, such virtual agents can help to enhance customer experiences, streamline operations, and provide personalized assistance.

- **Customer Service Support**: Generative AI can be used to create chatbots that can interact with customers in natural language and answer their queries about product specifications, functionality, or feedback and can even complete certain types of transactions. This is particularly useful as it can provide 24/7 instant support in multiple languages and reduce the workload on human customer service representatives. Target used AI chatbots to handle 10M customer queries in 2024, boosting satisfaction by 15%.

- **Customer Feedback Analysis**: Generative AI can summarize customer reviews and feedback across diverse sources and social media. It can highlight the strengths and shortcomings of the product. Businesses can use this information to enhance their products and refine marketing strategies. Customers can also make informed purchase decisions based on the feedback summaries. H&M used AI to analyze feedback and refining products for 5M customers in 2024.

Generative AI Adoption in Automotive

The automotive industry is poised to experience a significant shift powered by Generative AI as it improves fuel efficiency, driving comfort, and safety with advanced capabilities. Figure 3-8 shows some of the use cases in the automotive sector that are fueled by Generative AI.

Figure 3-8. *Generative AI Use Cases in Automotive*

- **Innovative and Efficient Vehicle Design**: Generative AI can be used to automate complex vehicle designs with optimized aerodynamic performance and improved fuel efficiency. It allows designers to experiment with different shapes and features. It can suggest more ergonomic, comfortable, and stylish interiors. It can also suggest raw materials with desired properties that can improve fuel efficiency or provide better heat resistance for engines. All this speeds up the development process, helping to launch new models to market faster and more cost-effectively, while ensuring quality and reliability. Toyota's Research Institute uses Generative AI to generate modern prototype designs for new electric vehicles (EVs) based on text inputs. General Motors is using Generative AI to design lighter and stronger parts for its vehicles.

- **Optimize Production Lines**: Generative AI can make the production processes more efficient and flexible. It allows car manufacturers to customize vehicles as per customer needs and preferences. It can also enhance quality control by inspecting vehicles and parts with higher precision using advanced algorithms to catch defects in the production line that humans might miss. By speeding up production, optimizing costs, and improving precision, Generative AI helps drive innovation in vehicle manufacturing. All of this leads to more satisfied customers and a competitive edge in the market.

- **Enhanced Customer Buying Experience**: Generative AI can transform the customer purchase experience in the automotive industry. By analyzing customer data, it can make personalized recommendations for vehicles, features, and financing options. Through virtual try-ons, Generative AI can help customers make informed and satisfying choices. It can streamline the overall vehicle purchase experience by automating paperwork, providing real-time assistance with customer queries, thereby reducing the decision-making time. This not only boosts sales but also builds a long-term customer relationship with a higher degree of customer satisfaction. For example, Ford used AI to personalize vehicle recommendations, increasing sales by 20% in 2024.

- **Autonomous Driving**: Generative AI can be used for real-time decision-making for autonomous vehicles. Trained on a large set of driving data, Generative AI can generate optimal responses and make decisions to apply brakes or accelerator or even change lanes to ensure a safe driving experience. Tesla is using Generative AI to simulate driving scenarios for the development of its Full Self-Driving (FSD) cars, reducing dependency on physical testing, and helping the neural network handle uncommon situations.

- **Personalized In-Car Experience**: Generative AI is transforming how drivers interact with vehicles, making in-car experiences more intuitive, personalized, and engaging. Automotive voice assistants in the vehicle can communicate with the driver in a natural language in an intuitive and context-aware manner, allowing users to set up navigation, play music, or send messages using the vehicle's infotainment system seamlessly. Mercedes-Benz, for instance, has integrated ChatGPT into over 900,000 vehicles in a beta program to provide advanced interactions and customizations for drivers. Based on user behavior, preferences, and historical interactions, AI can recommend music, podcasts, and other media that align with users' tastes. BMW is also developing a new driver assistance system using Generative AI for its "Neue Klasse" vehicles to enhance convenience and safety. Additionally, Generative AI can provide personalized navigation suggestions based on user preferences, driving history, and real-time traffic and road conditions to minimize delays.

- **Predictive Maintenance**: AI-driven predictive maintenance can help to ensure vehicle reliability and minimize breakdown. Based on the data gathered from various sensors and using machine learning models trained on past data, it can predict the possibility of any issues arising in future. While AI can make necessary recommendations for proactive interventions that can reduce the risk of unexpected breakdowns and costly repairs, Generative AI enhances this by simulating failure modes or generating synthetic sensor data to improve model training. Volvo used AI to predict maintenance, reducing downtime by 30% in 2024.

Generative AI Adoption in Travel and Hospitality

Generative AI in the travel and hospitality industry can be used to assist travelers during their booking process—provide personalized travel recommendations, prepare personalized itineraries, give a virtual tour for their travel destinations, and more. Travel companies can use Generative AI for dynamic pricing, automated customer service, targeted marketing campaigns, and for many more use cases. All of this can help travel companies to provide relevant content to their customers at reduced cost, increase speed to market, improve operational efficiency, and ultimately improve customer satisfaction and drive revenue growth.

Figure 3-9 shows some of the prominent Generative AI use cases in the travel industry.

Figure 3-9. *Generative AI Use Cases in Travel Industry*

- **Personalized Marketing**: AI algorithms can build a comprehensive profile for each customer with their interests in specific destinations, activities, travel durations, accommodation types, and budget based on their past bookings, search history, and interaction frequency. Travel companies can send personalized offers to their customers based on their preferences. Customers are likely to find such offers more relevant to their needs and decide to avail them immediately. For example, Booking.com used AI to personalize offers, boosting conversions by 25% in 2024.

- **Personalized Travel Recommendations**: Generative AI can identify preferences and vacation patterns for everyone based on their demographics and travel history and subsequently recommend holiday destinations and activities highly personalized for them and fitting into their budget. Such personalized recommendations not only make the search and booking process more engaging but also has a much higher probability of booking completion, resulting in higher conversions. It also fosters a positive brand perception for the company. According to AI Chat Platform Report(2024), Kayak has introduced an AI-powered chat platform that provides personalized options for destinations and accommodations aligned with user preferences.

- **Personalized Travel Itinerary**: Generative AI can create dynamic itineraries that are highly personalized to meet the individual's needs. Based on travel routes, connection for flights, including ticket prices and layover times, accommodation options, activities, and experiences and local transportation options, Generative AI can create a highly personalized travel itinerary for individuals fitting into their budget. Such a personalized itinerary helps to increase efficiency and boost the customer's travel experience. TripIt's Smart Itinerary Planner is an AI-driven travel companion that not only creates a comprehensive, day-by-day itinerary for the entire trip but also forwards and manages the booking confirmation vouchers for flights and hotels and provides maps to navigate destinations with ease.

- **AI-Powered Booking Support**: Chatbots powered by Generative AI can provide round-the-clock assistance to travelers. Priceline has integrated OpenAI's advantage voice mode technologies into its AI-powered chatbot to enhance customer experience with simplified voice-enabled search and travel booking. These chatbots can simplify the booking process by answering various queries related to flight alternatives, luggage fees, layover duration, total time and cost for the travel. Travelers can also get information and personalized recommendations on accommodation options based on travel behavior, preferences, and budget by interacting with the chatbot in their natural language. Such an experience not only delights the customer but also helps to improve overall productivity by using chatbots to perform repetitive operations. Expedia's AI-powered chatbot provides constant and personalized suggestions to customers and helps them to book hotels and flights, thus enriching the overall customer experience.

- **Dynamic Pricing Optimization**: With dynamic pricing, travel companies can modify their offers based on supply/demand cycle and other external factors like weather, location, local events, etc. Generative AI can analyze various factors and forecast future demands and recommend an optimal price for room rents for a given location and date. Travel companies can offer better deals and improve overall customer satisfaction by strategically lowering prices during off-peak times. Marriott International uses AI to dynamically adjust the room tariff based on demand and local events.

- **Language Translation Services**: Going to places where people do not speak the language of the traveler can be a challenging experience for travelers. Language translation services embedded in travel apps can allow travelers to navigate foreign destinations, interact with locals, understand specific signs, and local customs. AI-powered translation tools in apps and websites can immediately translate descriptions for hotels, tour packages, destinations along with booking instructions into preferred language for the customer. Tripadvisor used AI translations to engage 10M global users in 2024.

- **Real-Time Travel Risk Assessment and Advisory**: Generative AI can scan government travel advisory, weather reports, and social media for emerging risks due to political unrest, health outbreaks, and natural disasters. It can keep up with updates on hotel bookings and flight times and act as a virtual guide and suggest contingency plans in case of flight cancellations or any unexpected events. It can also recommend safer and optimized travel routes by analyzing traffic and weather conditions and even making necessary changes to their itinerary to make the best use of time and the situation. All this can make the travelers feel safe and secure during their travels. United AI uses AI models that provide timely information about flight disruptions and delays. Generative AI can also help travelers understand and comply with local travel restrictions, visa requirements, and health protocols.

- **Enhanced Traveler Interaction**: Generative AI can create high quality images and videos and craft compelling narratives about destinations. Such narratives can inspire and favorably influence the decisions of the travelers. Travel companies can create virtual travel experiences for their customers at a fraction of cost and time using Generative AI and sell the best possible holiday experiences. Expedia used AI visuals to inspire 5M travelers in 2024.

Generative AI is also transforming the hospitality industry by streamlining processes and enhancing guest experiences at all stages. From personalized pre-arrival communications to efficient check-in process, dynamic in-stay services, and tailored post-stay re-engagement, Generative AI is redefining luxury hospitality with enriching guest experiences, while also preserving the human touch. Figure 3-10 shows some prominent use cases for Generative AI in the hospitality industry.

Figure 3-10. *Generative AI Use Cases in Hospitality Industry*

- **Guest Reservation Management**: AI-powered virtual agents can handle booking enquiries and check for availability. It can provide personalized recommendations based on guest preferences, their period of stay, and loyalty status. Generative AI can be used to generate and send booking confirmations and reminders to the guests. It can also generate automated responses and updates for changes in reservations. Marriott has deployed a ChatGPT-Style AI assistant to assist guests with bookings, loyalty-based recommendations, and reservation modifications through mobile and digital channels.

- **Arrival Coordination**: Generative AI can record guest information, handle check-ins and room assignments on arrival. It can suggest upgrade options based on guest preferences and current availability. It can generate personalized welcome letters for guests and recommend local attractions, dining options, and activities tailored to guest preferences. On check-out, it can also generate personalized check-out summaries and invoices. Hilton uses AI-driven personalization to suggest room upgrades, enable mobile check-in, and tailor arrival experiences through its Hilton Honors app.

- **Housekeeping Management**: Generative AI can generate detailed checklists and automated assessments for room readiness and cleanliness. It can generate alerts for housekeeping with room status and specific guest needs. Optimized daily cleaning schedules based on room occupancy and guest preferences can also be created by Generative AI. It can also analyze guest feedback and identify areas of improvement. Marriott uses AI to prioritize room cleaning based on occupancy, guest preferences, and check-in patterns, improving turnaround time.

- **Guest Service Request Handling**: Generative AI can automatically respond to guest requests from mobile apps, in-room tablets, or voice assistants based on their preferences to provide 24/7 support. It can route the requests to the right department and assign tasks to staff based on workload, guest needs, and priorities. It can also track the requests in real-time and provide intermediate updates via SMS or app-notifications or in-room screens. In case of any potential delays, it can notify guests and suggest adjustments. Generative AI can also analyze guest feedback and identify areas for improvements and offer personalized compensation for service delays to maintain guest satisfaction. Rosewood hotels use AI-driven guest messaging platforms to respond instantly to service requests and proactively resolve issues, thereby reducing the service response time.

- **Departure Coordination**: Generative AI can generate invoices using predefined templates during check-out, reducing manual errors. It can recommend the best payment methods to guests based on their preferences and transaction history. Additionally, it can actively collect feedback from guests in real-time, ensuring timely and relevant responses. By analyzing reviews and feedback, it can identify areas for improvement and suggest practical changes to enhance the guest experience. Accor hotels have deployed a real-time guest feedback system to collect and analyze feedback at checkout and trigger immediate service recovery actions when needed.

- **Predictive Maintenance**: AI can analyze data from sensors, service logs, and guest feedback to predict equipment failures (e.g., HVAC issues, plumbing leaks) and schedule preventive maintenance before problems arise. It can send alerts to service teams or raise service tickets before guests report any complaints. It can help hotels stay compliant with safety regulations by predicting when inspections or updates are required. Hilton uses sensor data and AI analytics to predict HVAC failures and plumbing issues before they impact guests.

Generative AI powered by data will enable hotel managers to predict their customer needs better and will enhance human connection between hotel staff and their guests through personalized interactions—such as greeting a returning guest by name, remembering their room or pillow preference, proactively arranging a vegetarian meal, or suggesting activities aligned with their interests. While there are challenges around the implementation costs of AI-based solutions and aligning AI to the brand values of the hotel, the future of Generative AI in transforming the hospitality industry is promising. Generative AI holds a strong potential to bring in significant transformation in both front-of-house interactions and back-office operations. The hotel industry must also prepare this workforce to embrace this technological shift to keep the right balance between technology and human touch. Navigating the challenges of AI adoption to provide exceptional guest experiences and operational excellence needs a thoughtful and strategic approach.

Generative AI Adoption in Education

Generative AI has the potential to revolutionize how educational institutions approach teaching and learning. It has huge potential and when used rightly, Generative AI can personalize learning and improve educational outcomes. Figure 3-11 shows some of the use cases of Generative AI in learning and educational development.

Course Design and
Content Creation

Personalized
Learning & Tutoring

Automated
Grading

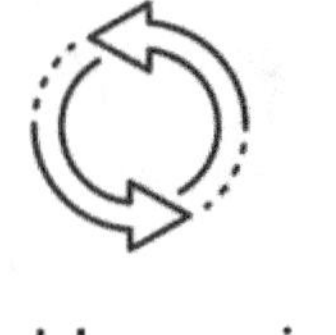

Old Learning
Material Restoration

Figure 3-11. *Generative AI Use Cases in Education*

- **Course Design and Content Creation**: Generative
  AI can design and organize course materials, lesson
  plans, and assessments for various subjects based
  on the student's knowledge gap and learning style. It
  can also generate new teaching materials, including
  explanation, summaries, quizzes, and questions for
  evaluation, and study aids like flash cards and guides.
  Additionally, it can produce images for a target topic to
  assist teaching. Harvard uses Generative AI to prepare
  curricula, create lesson plans and summarize course
  content, and more for teaching.

- **Automated Grading**: Generative AI can be used to
  assess and grade student assignments, tests, quizzes,
  and more. It can provide constructive feedback to
  students more efficiently. It provides valuable insights

to identify trends and gaps in learning. Overall, this saves time for teachers, allowing them to focus more on other critical aspects of teaching and learning.

- **Old Learning Material Restoration**: Generative AI can be used to restore low quality learning materials such as historical documents and photographs. Improving the resolution of old historical images helps students to read, analyze, and understand the material more easily, leading to better learning outcomes.

- **Personalized Learning and Tutoring**: Generative AI can be used to create personalized learning plans and virtual tutors for students. It can customize lessons to each student's pace and skill level, helping them to learn in their own way. This helps students develop a deep understanding of concepts, without feeling rushed. Quizlet's AI-powered Q-Chat generates practice questions, explains answers, and quizzes students in a personalized way. An AI-powered virtual tutor can interact with students and provide real-time feedback and support. It is helpful for students without access to in-person tutors and is a step toward providing good-quality education for all. Khan Academy launched *Khanmigo,* an OpenAI-powered virtual tutor and teacher assistant that provides an engaging and ethical approach to help students learn on various topics in a methodical and interactive manner. The flexibility and personalization enhance the overall learning experience and also help to democratize education for underserved regions.

Generative AI Adoption in Energy and Utilities

Like in all other industries, the energy industry—encompassing sectors like oil and gas, renewable energy and utilities, is looking at Generative AI to optimize operations, increase efficiency, and reduce costs. Figure 3-12 shows some of the use cases where Generative AI is being used in the energy and utilities industry.

Figure 3-12. *Generative AI Use Cases in Energy Industry*

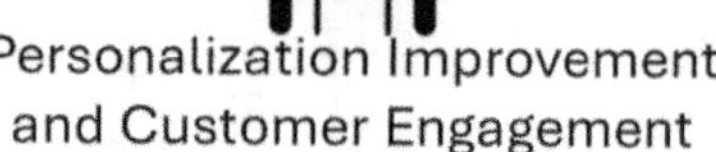

Figure 3-13. *Generative AI Use Cases in Utilities Industry*

Use Cases for Energy Industry

- **Reservoir Simulation and Modeling**: Generative AI can generate accurate models of underground oil and gas reservoirs. This helps to make informed decisions about drilling and production strategies.

- **Drilling Optimization**: Generative AI can be used to optimize drilling operations by analyzing geological data, drilling history, and sensor data coming from various sources like geological surveys, drilling operations, End of Well reports, and internal IT systems. It can generate optimized drilling plans that can minimize cost and maximize energy efficiency. It can identify wells with significantly higher non-production time (NPT) due to equipment failures, repair, or the equipment is either lost or stuck in the wellbore.

- **Production Forecasting**: Generative AI can be used to analyze historical climate and weather data, satellite images, IoT sensor data, and provide a precise production forecast for renewable energy like solar, hydro, and wind, which are often variable and intermittent. Based on the production forecast, grid operators can balance supply and demand. They can also adjust the grid operations accordingly to maintain stability and reliability. This can also help in resource allocation and investment decisions to meet future demands.

- **Environmental Impact Analysis**: Generative AI can be used to analyze emissions and environmental data and help to foster sustainable practices. It can help companies optimize their carbon reduction strategies and be compliant with ecological regulations. It can be used to identify optimal locations and configurations for installing solar panels and batteries that can reduce energy wastage.

- **Safety and Sustainability Optimization**: Generative AI can analyze digital images from cameras stationed in remote and hazardous environments for potential safety risks. It can detect gas leaks from faulty valves and recommend protective equipment and procedures for remedial work.

Use cases for Utilities Industry

- **Grid Performance Optimization**: Generative AI can analyze data from smart meters, grid devices, and other sources and predict voltage, frequency, power factor, and congestion of the electrical grid. It can monitor and control the flow and quality of electricity and ensure that the grid remains stable, reliable, and resilient. This improves operational efficiency and flexibility and reduces grid losses and emissions.

- **Supply Chain Optimization:** Generative AI can predict future demands and generate optimized supply chain strategies by analyzing historical demand patterns, environment data, transportation costs, and inventory levels. This helps energy companies to reduce costs, improve efficiency, and ensure on-time delivery of resources.

- **Predictive and Preventive Maintenance:** Generative AI can be used to proactively identify faults, anomalies, and failures in equipment and systems. It can analyze the data from sensors, cameras, drones, etc. and detect signs of potential problems. Based on the identified problems, it can recommend corrective actions and maintenance requirements, before it leads to shut down or total equipment failure. Such preventive capabilities can prevent delays, reduce downtime, and avoid expensive repair costs.

- **Routine Tasks Automation:** Generative AI can be used by utility companies to automate routine tasks like data entry, invoice generation, payments and document processing, and other operational tasks. This can help to improve efficiency and productivity by eliminating repetitive and time-consuming tasks that are error prone.

- **Personalization Improvement and Customer Engagement**: Utilities companies can use Generative AI to analyze customer data, understand their preferences, and respond in a personalized manner to their needs and queries via chatbots in natural languages or through personalized emails. Generative AI powered chatbots can engage with customers in natural language and provide guidance on their problems and queries in a personalized manner.

- **Market Trading and Pricing Optimization**: Generative AI can help to forecast energy prices based on renewable supply fluctuations. Based on that, energy providers can make informed trading decisions in power markets.

Key Industry Takeaways

Table 3-1. *Key Industry Use Cases and Primary Benefits*

Industry	Key Generative AI Use Cases	Primary Business Value
Manufacturing	Product Design, Predictive Maintenance, Supply Chain Optimization	Cost reduction and operational efficiency
Healthcare & Life Sciences	Medical Imaging, Drug Discovery, Personalized Medicine	Improved patient outcomes and accelerated innovation
Banking	Customer Service, Fraud Detection, KYC Automation	Better customer experience and risk management

(continued)

Table 3-1. *(continued)*

Industry	Key Generative AI Use Cases	Primary Business Value
Insurance	Claims Automation, Underwriting Copilots, Fraud Detection	Faster claims processing and lower operational costs
Capital Markets	Research Automation, Wealth Management, Algorithmic Trading	Improved decision-making and productivity
Media & Entertainment	Content Creation, Video Generation, Personalization	Faster content production and audience engagement
Retail & eCommerce	Personalization, Demand Forecasting, Virtual Shopping Assistants	Revenue growth and customer loyalty
Automotive	Vehicle Design, Autonomous Driving, Predictive Maintenance	Innovation, safety, and efficiency
Travel & Hospitality	Personalized Itineraries, Dynamic Pricing, Virtual Assistants	Enhanced customer experience and revenue optimization

Across industries, Generative AI has the potential to deliver three main benefits – boosting operational efficiency, improving customer experiences, and speeding up innovation. The organizations that successfully adopt Generative AI and embed AI into their core business processes will emerge as winners.

Conclusion

Generative AI is more than technological enhancements. It has moved beyond the experimentation phase to become a transformative force reshaping industries by unlocking unprecedented efficiencies, personalizing customer experiences, and driving innovation at scale. Businesses across all industries are using Generative AI to not only solve complex business problems but also deliver measurable impact. Organizations are already reporting **20–40% productivity gains, double-digit cost reductions**, and **faster time-to-marke**t.

While the potential is huge, businesses must navigate ethical concerns - **data privacy, bias mitigation, and intellectual property risks**. As discussed in Chapter 2, a robust governance framework, human oversight, and transparency will be critical to ethical and sustainable AI adoption. While the capabilities of Generative AI are fast improving, the question for businesses is no longer if but how to adopt it strategically. Business executives need to focus on building an equitable global economy with those use cases that drive ROI and growth ethically and responsibly and give a competitive advantage with reduced risks.

CHAPTER 4

Assessing Your Generative AI Readiness

Introduction

Generative AI holds the promise of unmatched efficiency, creativity, and innovation. But without readiness, that promise can turn into risk. Before moving Generative AI from experimentation to mainstream production, executive leaders and C-level executives need to assess the organization's readiness for company-wide adoption of Generative AI projects. Success depends on having the right strategy, trusted and governed data, scalable technology, skilled teams, strong governance, and disciplined financial planning.

This chapter will guide you to assess and strategically align Generative AI projects to business goals. It will explain how to check for data availability and quality to ensure better outcomes and reliability. We will look at how to select the right technology and Generative AI models, and how to assess the scalability and availability of the current technology infrastructure to meet customer needs and expectations. We will also learn to assess the cultural and workforce alignment for talents and skills,

© Brajesh De 2026

B. De, *Generative AI for Business Innovation*, https://doi.org/10.1007/979-8-8688-2682-5_4

operational readiness, and new governance needed to build responsible and ethical AI solutions. Finally, this chapter looks at the financial and ROI assessment, without which the overall assessment for Generative AI adoption is incomplete. Overall, this chapter provides an actionable framework to assess where you stand, identify gaps, and build a clear roadmap from pilot to production.

Why Is Generative AI Readiness Important?

The Reality: According to a report published by MIT, 95% of Generative AI pilots are failing, often due to poor data quality, unclear strategy, and lack of organizational alignment.

Lessons Learned: The success of Generative AI pilots depends more on the foundations than on the model. Most of them fail due to poor data quality, unclear business objectives, and lack of organizational alignment.

The Reality: The University of Texas M.D. Anderson Cancer Center spent millions of dollars to build an AI system powered by IBM Watson that provided erroneous cancer treatment advice, largely because it was trained on a small, hypothetical dataset rather than real-world patient data.

Lessons Learned: Generative AI systems must be trained on realistic and grounded datasets. Their outcomes must be continuously validated by domain experts, especially in industries that have high stakes.

The Reality: US sports publication "Sports Illustrated" was embroiled in a scandal after publishing AI-generated articles written by fake authors.

Lessons Learned: There are reputational risks of using Generative AI without transparency. Organizations must disclose the use of Generative AI and enforce content governance to preserve trust with society.

The Reality: A large North American airline was held responsible for its AI chatbot providing a customer with incorrect information about bereavement fare refunds due to a lack of human oversight.

Lessons Learned: Human oversight is essential to bridge the gaps and validate AI-generated responses. Clear escalation path must be established wherever humans are interacting with systems powered by Generative AI.

All these are real examples where rushing into Generative AI adoption without assessing the readiness resulted in notable failures. They highlight the importance of doing a readiness assessment before adopting Generative AI technology, which can be unpredictable, costly, and ethically complex.

Readiness assessment reduces risks, accelerates time-to-value, and maximizes ROI. Prioritizing Generative AI readiness can help to drive innovation, reduce cost, and improve revenue. While these promises are enticing, deploying Generative AI technologies successfully requires more than just technical know-how. Businesses need to take a well-rounded, holistic approach to be ready from various dimensions to embrace and build new innovative solutions using Generative AI. Hence, an assessment is generally the very first right step in the journey for Generative AI adoption.

The Eight Pillars of Generative AI Readiness

Organizations often focus on models and technologies to assess their AI readiness. However, a successful Generative AI adoption needs a balanced evaluation across multiple dimensions. The following eight foundational pillars provide a practical framework to assess organizational capabilities needed to ensure the success of Generative AI implementations. Organizations must assess their capabilities around these pillars to know their Generative AI readiness:

1. **Strategic Alignment to Business Goals**

   Generative AI investments must be directly tied to measurable business outcomes. When aligned with strategic business objectives like revenue growth, cost reduction, productivity gains, customer satisfaction, or risk mitigation—AI delivers tangible value. Clear ownership and well-defined success metrics help to keep AI initiatives aligned to value-driven outcomes and ensure they do not become another experiment.

2. **Data Quality and Availability**

   Data is the foundation for successful Generative AI. A successful Generative AI implementation needs high-quality and on-time data. Organizations must have efficient data pipelines to ensure that data is accurate, complete, representative, secure, trustworthy, and unbiased for Generative AI to deliver business value. Poor data quality has been one of the leading reasons for the failure of Generative AI.

3. **Technology Infrastructure for Scalability**

Without a scalable and secure technology infrastructure, Generative AI pilots often fail to scale when deployed in production. The infrastructure must also be cost-efficient to scale with the growing traffic. A cloud native architecture built using APIs and Microservices deployed on a cloud platform provides the right technology foundations, especially when clubbed with robust and flexible LLMOps, Infrastructure decisions made during the pilot implementation decide if the solution can scale sustainably.

4. **Model Evaluation and Selection**

Choosing the right LLM model is a strategic decision that affects accuracy, latency, and cost. One model that fits all may not be the practical solution. Hence, organizations must have a disciplined framework to judiciously select the right LLM model from the list of proprietary and open-source models. Without it, organizations risk selecting an incorrect model that is not optimized to their needs and may incur both reputation and financial damage.

5. **AI Governance for Legal, Ethical and Responsible AI**

Strong AI governance is a business imperative. It ensures regulatory compliance, protects intellectual property, ensures safety, mitigates biases, and promotes transparency. By doing so, it builds trust, prevents costly legal or reputational fallouts, and fosters innovation with societal values.

AI governance must establish clear policies for data usage, auditability, accountability, and human oversight for sustainable adoption.

6. **Talent, Skills, and Organizational Culture**

Talent, skills, and organization culture are fundamental for the successful adoption of Generative AI and maximize its benefits. Success depends on collaboration of multi-disciplinary teams involving business and technology leaders, domain experts, data scientists, prompt engineers, and legal professionals. Organization culture must support responsible experimentation, continuous learning, and change adoption.

7. **Financial Readiness**

Financial readiness is important to **mitigate significant operational, regulatory, and financial risks**. Organizations must plan for the substantial cost and investments required for high-performance computing infrastructure, model usage, data quality and governance, and ongoing operations. They must also account for talent acquisition and upskilling and manage high-stake risks due to data privacy violation, AI hallucinations, and biases.

8. **Pilot and Feasibility Testing**

Pilot implementation and feasibility testing is crucial for organizations to evaluate the technical viability, ethical implications, and return on investment (ROI) of Generative AI solutions. Pilots help to validate that the Generative AI solution can offer real business value aligned to the organization's specific needs. It minimizes the risk of expensive, large-scale failures and provides a data-driven foundation for confident and strategic Generative AI adoption.

Key Takeaway: *Generative AI readiness is not a technology checklist—it is a coordinated investment across strategy, data, platforms, governance, people, and finance to enable responsible, scalable impact.*

Figure 4-1 shows the framework that must be considered for assessing Generative AI readiness.

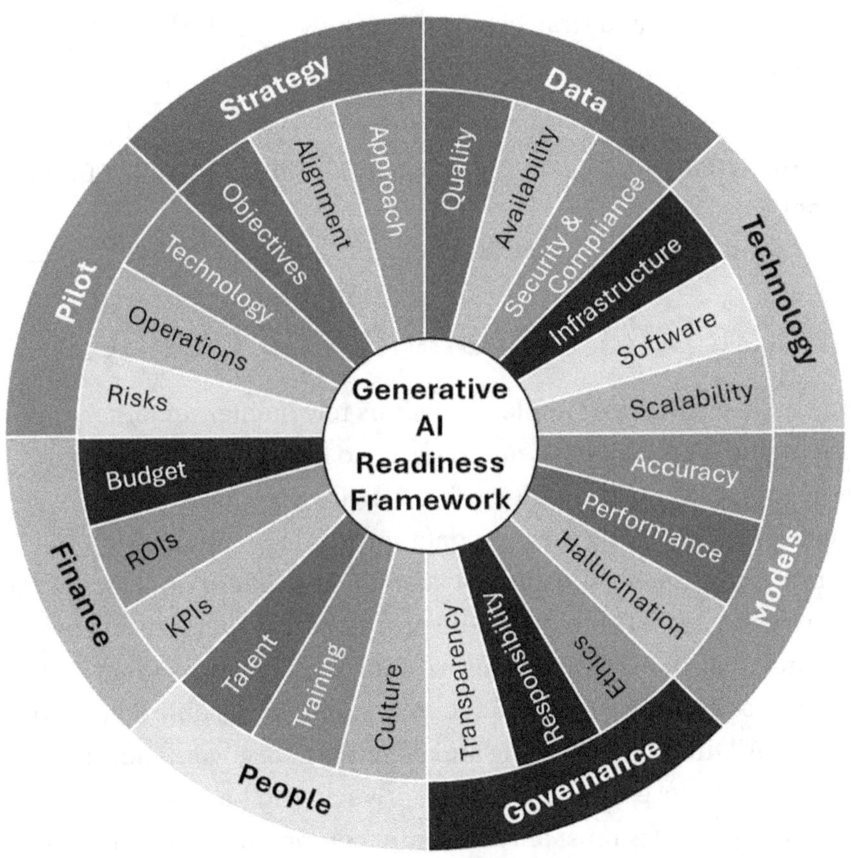

Figure 4-1. Generative AI Readiness Framework

Assess Strategic Business Alignment

For a successful adoption and implementation of Generative AI, there should be a strategic alignment of the Generative AI plan with the overall business goals. The leaders and partners of the organization must also be ambitious and strategically aligned and support the Generative AI initiatives. There must be clear objectives that define why and how Generative AI will be implemented. There must be an approach with clear policies and guidelines for using Generative AI. Organizations should promote their AI thought leadership through research and external community engagement. Stakeholders within the organization should agree on our AI-related strategy, including how AI tools should be used and deployed.

Pro Tip   Treat AI adoption like a business transformation, not an IT project.

Evaluate Data Availability and Quality

Successful Generative AI implementations rely on high-quality interpretable data. An organization needs to have a data strategy and system in place to collect high-quality data from various sources. This involves identifying the right data sources and entities and understanding the relationship between them to better understand the business processes. There should be processes and methods in place to continuously monitor and improve data quality and accuracy over time. Data collected needs to be securely stored in scalable data storage solutions. All this needs a strong and effective data governance framework to manage data access, quality, usage, privacy, and security. The data governance practices must always ensure compliance with data privacy regulations for the industry and region.

Most organizations have multiple data sources that may be siloed. A unified data fabric—breaking down the silos—is needed to ensure accessibility and context. All different data sources should be integrated for seamless and secure access and flow of data across teams, systems, and platforms. Organizations must leverage technologies and implement best practices to secure data from various security vulnerabilities. Regular assessments and testing to ensure the effectiveness of data security measures are important.

Ensuring the quality of the collected data will form the foundation for high-quality Generative AI output. Training AI models with good and high-quality data will help the model to learn the relationships between different datasets and make accurate and reliable predictions. Some of the main quality attributes of data are completeness, accuracy, consistency, validity, and timeliness. Outliers in the data should be addressed before being used for any model training.

Once the necessary and right data has been collected, it is important to derive insights from it to power Generative AI initiatives. Real-time data flow and orchestration are needed so that insights, alerts, and decisions can happen as events unfold. But processing data in real time and deriving immediate insights is complex. A strong foundation in data analysis with the right tools is needed to derive valuable insights to power Generative AI applications.

Automated data management processes with minimal human intervention are needed to enhance the overall quality of data. Advanced techniques for data versioning and monitoring are needed to ensure the accuracy and consistency of data. A comprehensive data backup and disaster recovery plan that is regularly tested can enhance the data protection needed for AI systems.

Overall, strong data management practices are necessary to ensure the highest levels of quality and compliance needed for Generative AI.

Pitfall: Siloed or poor-quality data can lead to biased, unreliable AI outputs and undermine Generative AI initiatives, no matter how advanced the AI models are.

Technology Infrastructure Scalability Assessment

Once an organization selects the use case to adopt Generative AI, it needs a trusted platform to implement and run it. Generative AI applications require specialized infrastructure to support their computational and memory needs for processing large datasets and running real-time analytics. A smooth integration of AI tools with existing IT infrastructure and scalability to accommodate future growth is paramount. A technology infrastructure assessment is important to predict the success of Generative AI initiatives.

Without a scalable technology infrastructure, Generative AI initiatives will fail to meet future demands without compromising performance, cost, security, or user experience.

Generative AI applications need high computational power, GPUs/TPUs, and data storage that are generally available in the cloud. Hence, organizations must be cloud-ready and host their applications on a cloud infrastructure.

There is no AI without API. This is driving an exponential growth of APIs that underpin all Generative AI models. Hence, an API management platform is essential to manage the expanding API landscape. The traffic management and load balancing capabilities of these API management platforms are particularly useful for AI implementations.

DevSecOps practices play an important role in managing smooth deployment of AI models and prompts. Existing CI/CD pipelines should be modified to validate Generative AI responses that are not always the same, unlike traditional applications. MLOps is yet another technology addition needed to manage the life cycle of AI models that may have to be retrained continuously.

Generative AI applications may need to be specially secured to prevent misuse. Hence, authentication protocols and mechanisms for controlling access should be reviewed and enhanced, if required. Secure key management, caching, and multi-region deployment are some of the other technical aspects that must be reviewed as part of a Generative AI technology infrastructure assessment.

Quick Win: Leverage existing cloud investments and partnerships to accelerate and build a robust, cloud-ready, and secure technology infrastructure to successfully deploy and manage Generative AI applications.

Evaluate Generative AI Models

There are several Generative AI models out in the market today, both open source and closed source. Each of them excels in different areas. Hence, while evaluating Generative AI models, the primary use case should be the first criterion since different applications demand different evaluation approaches. The use cases can be chatbots, summarization, code/text generation, or any other content creation.

After finalizing the use cases, look at the metrics that can impact the outcomes—importantly, accuracy, speed, performance, and cost savings. Other considerations that should be part of the model evaluation are biases, hallucinations, and data privacy. There are quantitative and qualitative metrics that must be combined. The metrics and assessment methods must be aligned with both business goals and user expectations.

The performance and accuracy of the model for text generation can be assessed through various quantitative metrics like BLEU, ROUGE, METEOR, Perplexity, and Self-BLEU. Similarly, Inception Score (IS),

Fréchet Inception Distance (FID), and Structural Similarity Index (SSIM) can be used to measure the accuracy of image generation. Truthfulness and faithfulness can be used for hallucination checks. The qualitative metrics should involve human evaluators to judge the output for content creativity and novelty, relevance and appropriateness, coherence and consistency, accuracy, and factual correctness. Business leaders should focus on whether the model delivers accurate, safe, compliant and cost effective outcomes for their intended use cases.

Robustness and safety checks for models should also be part of the evaluation parameters. Include model performance and assessment for adversarial prompt testing for "jailbreak." How the model performs in response to malicious prompts will determine its robustness. Check the model output for bias across demographics, gender, and religion to ensure fairness. Last but not least, ensure that the model output conforms to regional and regulatory compliances like GDPR, CCPA, and industry-specific rules.

Decision Rule: Always select the model that best fits the business use case and optimizes performance, reliability, compliance, and practical deployment needs. Balance accuracy, safety, cost, and regulatory considerations while selecting the right model.

Legal and Ethical Considerations

Legal and Ethical considerations are needed to protect the brand, customers, and financial interests of the organizations. It protects the organization from legal liability and regulatory risks. It safeguards the brand reputation and customer trust. It ensures responsible data usage and protects intellectual property. Hence, all Generative AI applications must be developed and deployed following strong ethical principles.

Organizations must establish clear guidelines and policies in place that promote the responsible use of Generative AI systems. They should strictly enforce all guidelines and review them regularly to ensure they remain relevant and effective.

A key principle is **fairness.** Generative AI systems should not amplify any kind of biases. Outputs should be impartial and accurate. AI decision-making should be transparent, with technical and comprehensive explanations of the results produced. All required information should be available to the user based on the permissions granted. Maintaining transparency helps to build trust with stakeholders and enhances accountability. Comprehensive frameworks must be in place to ensure that AI systems are transparent, explainable, and accountable for their actions.

Additionally, Generative AI systems must **comply with the relevant laws and regulations** of the industry and region. This includes adhering to data privacy laws, such as GDPR, which protect user information and define how data should be handled. Staying aware and informed about the laws and regulations protects the organization from legal risks. Best practices must be adhered to, and compliance ensured at all times. There should be a comprehensive set of safeguards in place to prevent misuse of Generative AI. All this can help build a responsible foundation for the implementation of Generative AI that respects both legal and ethical boundaries.

User content is critical when collecting any data that can be used for building AI systems. All AI systems should always operate under human oversight to prevent unintended consequences. Employees must undergo comprehensive training on ethical considerations in AI. That will establish a culture of ethical decision-making throughout the organization.

Risk Radar: Bias amplification, hallucination, model misuse, and data privacy violations are the core threats that can derail a responsible AI strategy without human oversight.

Pillars of Talent and Skill Readiness

Technical Fluency

Generative AI development needs special skills to work on relevant AI/ML and Data science technologies, architecture, platform, and tools. It needs knowledge about different Generative AI algorithms and frameworks. Some of the most common Generative AI algorithms are Variational Autoencoders (VAEs), Generative Adversarial Networks (GANs), and autoregressive models. Popular frameworks for building Generative AI applications include TensorFlow, PyTorch, Keras, LangChain, LlamaIndex, and Hugging Face Transformers. Organizations need to assess their experience and knowledge with these frameworks to build fully functional and scalable Generative AI applications.

Continuous Learning

The technologies for Generative AI are evolving. Hence, the workforce must also keep up with the rapid evolution of Generative AI. They will need to evolve and learn new skills, iteratively and over a period of years.

Leadership Awareness

Leaders must understand how and where Generative AI can create value instead of just having technical knowledge about it. They must also keep track of the key developments in emerging Generative AI technologies. Executives need to be clear and pragmatic about the opportunities, capabilities, and risks of Generative AI.

Sourcing Strategy

Knowing the skill gaps helps to plan the upskilling and training needed to ensure that the team possesses the necessary skills before moving forward. Some gaps may be filled by internal training. It may need to partner with educational institutes, online courses, webinars, or workshops. But in some cases, hiring external consultants to strengthen the team can accelerate the process and bridge the skill gaps. An assessment to balance the internal training with external talent acquisition can help the team and the firm prepare for a smooth transition to adopt Generative AI technologies.

Training strategy: Build a culture of continuous learning for the evolving Generative AI landscape with a mix of internal training and strategy hires.

Financial Readiness

Generative AI is not a one-time technology purchase. It is an ongoing financial commitment that spans data, infrastructure, models, talent, integration, governance, experimentation, and scaling. Rigorous financial planning is needed to ensure the success of Generative AI initiative - without that, it risks stalled pilots, cost overruns, and failure to transition from proof-of-concept to measurable business value.

Cost Allocation

Most organizations have been playing around with Generative AI technologies to build proof-of-concepts and pilot solutions. True and significant ROI can be achieved only when they scale these solutions and take them to production. But that needs further investments to build advanced solutions. Assessing financial readiness is important to ensure a sufficient budget to successfully implement and scale Generative AI solutions.

Organizations need to have a clearly defined budget for Generative AI, covering experimentation, PoCs, scaling, and operationalization. They need to plan ahead for specialized infrastructure needed to build Generative AI solutions and for ongoing cloud and AI services. Token-based pricing for LLMs can scale unexpectedly. Organizations must ensure investments in data lakes, pipelines, and governance frameworks needed for RAG or fine-tuning scenarios. Budget must also be allocated for cybersecurity and regulatory compliance tools specific to AI. Hence, they need to plan for initial CapEx and ongoing OpEx for all Generative AI initiatives. Organizations must have a good understanding of pricing models for cloud and AI services and should closely monitor every expenditure to maintain budget control.

Along with infrastructure, organizations must factor in the cost of training existing staff or hiring new talent with expertise and skills in Generative AI, MLOps, and prompt engineering. Cost for consulting, integration, and support services from AI vendors or system integrators must also be budgeted.

The cost of scaling the pilot implementation for production deployment can be significant and must be part of financial readiness planning. Additionally, the costs of continuous evaluation, prompt updates, fine-tuning, and retraining models must also be factored into the overall budget.

Last but not least, allocate contingency budgets to handle failed experiments or unexpected costs. Also, account for potential penalties or legal costs if AI adoption is non-compliant. 10 to 15% must be budgeted for failure buffers. As a best practice, tying financial KPIs to Generative AI initiative goals can drive executive buy-in and ensure financial readiness for a successful Generative AI implementation.

Measuring ROI

Measuring and justifying the ROI of investments in Generative AI is one of the biggest challenges for CFOs. If the investment in Generative AI cannot translate to measurable ROI, success remains undermined. For an AI investment, the ROI can be financial (direct revenue) and operational (productivity improvements). The key ROI metrics are cost savings, revenue uplift, and productivity gains. Measuring the true ROI from Generative AI can be a challenge due to the following:

- To realize satisfactory ROI from Generative AI can be a time-consuming process—often takes 2 to 4 years.

- It is hard to separate the gains from AI initiatives from those of other initiatives, like operational excellence, team reorganization, or changing roles.

- Fragmented systems and siloed platforms make it challenging to track before-and-after impact.

- Cultural resistance, employee adoption, and workflow adaptation for Generative AI can slow down the value realization.

Lack of a well-defined framework to calculate the ROI, can derail further investment plans for Generative AI. The framework for measuring the ROI from Generative AI must consider a wider set of KPIs, including cost savings, revenue growth, productivity improvements, customer satisfaction, and risk management. Organizations must also continuously track the performance of Generative AI initiatives and adjust enterprise strategy as needed to maximize ROI.

Feasibility Assessment and Pilot Testing

Pilot projects and prototypes help validate the Generative AI use cases before fully committing resources for mainstream adoption. Start Generative AI implementations with a small prototype with a limited scope. The objective should be to assess the technical feasibility and the business impact. This helps to identify potential issues and understand the real-world impact without significant risk.

The pilot should be small enough to have low operational risk but comprehensive enough to give clear evidence of value, risks, and next steps. There should be a visible business impact with measurable output if it becomes successful. Performance metrics like speed, accuracy, time saved, and cost reduction should be evaluated. Some qualitative KPIs to measure the effectiveness of the pilot can be user satisfaction, ease of adoption, and creativity improvement. Data constraints and privacy requirements should also be part of the feasibility assessment.

The feasibility assessment from the pilot implementation must cover the following:

- **Technical Feasibility**: It should cover the accuracy of the model, relevance for the domain, latency, scalability, and integration complexity with existing systems.

- **Operational Feasibility**: It should cover the changes that may be required in the existing systems and process workflow. New skills may be required to maintain and enhance the solution. Any impact on security posture and compliance implications must also be assessed.

- **Economic Feasibility**: A cost-benefit analysis should be done to evaluate the potential ROI. Additional costs may be required for ongoing inference and model fine-tuning.

- **Risk Feasibility**: Risks due to hallucination, bias, and data leaks should also be evaluated.

Test the pilot with real-world scenarios in a controlled environment. Start with a small group of users to test the pilot. Capture the completion time, output quality, error rates, and user feedback to measure the success of the pilot implementation. Evaluate the results of pilot testing against the success criteria to decide whether to scale, pivot, or stop. Accuracy vs. cost, speed vs. security, and automation vs. oversight are decision levers in a Generative AI feasibility and scaling discussion. They help in balancing priorities when moving from pilot to production by making explicit which aspects are more valuable, because improving one usually impacts the other.

Conducting the Assessment

The Generative AI readiness assessment reduces risks and accelerates time to market for any application or feature built using Generative AI technologies. Generative AI poses unique challenges but also offers unprecedented opportunities. An assessment can guide the organization to reduce the entry barriers by selecting the right use case and identifying the execution path. It can help to address the common challenges with data, technology, infrastructure, and governance for a smooth Generative AI adoption.

Here is a structured approach and best practices for conducting the assessment:

1. Define the purpose and scope of the assessment for high impact AI outcome:

    a. Understand if the assessment is for a single use case or multiple pilots of an enterprise-wide adoption.

    b. Prioritize business areas where Generative AI delivers measurable value and drives tangible outcomes.

    c. Identify and engage the right stakeholders from business, data, technology, compliance, and security to accelerate adoption and alignment.

2. Assess Business Readiness to secure buy-in and demonstrate wins:

    a. Secure leadership buy-in and budget commitment to accelerate adoption and long-term impact.

    b. Start with low-risk, high-visibility pilots to demonstrate quick wins and build organizational momentum.

    c. Make certain that KPIs for success, like ROI, customer experience, and efficiency improvement, are defined to measure business outcomes.

3. Assess Data Readiness for improved model precision and reliability.

    a. Validate that the required domain-specific data is available for effective fine-tuning and model performance.

    b. Strengthen data quality by looking at various quality attributes like completeness, accuracy, consistency, and validity to improve model reliability.

   c. Implement governance policies that safeguard data security and ensure controlled accessibility for compliant AI operations.

   d. Maintain accurately labeled data to enhance training outcomes and model precision.

4. Review AI Model Strategy for safe, high-performance AI.

   a. Evaluate hosting options for different types of LLM models needed.

   b. Identify the appropriate model type needed for the specific use case.

   c. Establish effective processes for monitoring, evaluating, and tuning models for fairness, safety, performance, and compliance.

5. Assess Technology and Infrastructure Readiness for scalable and seamless integration.

   a. GPU/TPU resources must be provisioned for scalability.

   b. Establish MLOps/AIOps capabilities for monitoring, logging, and model retraining.

   c. Evaluate integration capabilities for RAG pipelines, embedding stores, and existing systems.

6. Assess Talent, Skills, and Organizational Readiness to accelerate AI initiatives.

   a. Have skilled and experienced prompt engineers in place.

   b. Include AI/ML Engineers, MLOps Engineers as part of the core team.

    c. Conduct workshops to upskill teams before the pilot.

    d. Secure cross-functional buy-in from IT, HR, and key business units.

7. Assess Governance, Ethics, and Compliance to establish trust on AI.

    a. Establish guidelines for safe and responsible AI policies.

    b. Define and implement detection and mitigation policies.

    c. Promote transparency and explainability of the Generative AI outcomes.

8. Conduct a pilot project to measure the value and prove business impact of AI.

    a. Build and test Generative AI with small-scale, low-risk prototypes aligned to prioritized use cases.

    b. Measure pilot outcomes using defined performance and business impact metrics.

9. Document outcomes and create a roadmap for enterprise-scale AI adoption.

    a. Aggregate insights from each assessment area into an executive report.

    b. Build a roadmap to address the identified gaps, improve readiness, and maximize the value of Generative AI.

    c. Iterate for the successful scaling of the pilot to production.

Conclusion

Realizing the full potential of Generative AI-driven innovation requires careful planning, preparation, and readiness assessment. By systematically evaluating key dimensions such as strategy, data quality, technology infrastructure, organizational capabilities, model selection, legal and ethical frameworks, talent and skills, and financial preparedness, organizations can enhance their readiness to harness the power of Generative AI effectively. A thorough readiness assessment enables organizations to identify opportunities, mitigate risks, and ensure scalable growth. With a robust assessment approach, organizations can confidently transition from experimentation to real-world implementation, positioning themselves for sustainable success in the Generative AI era. Organizations that invest in readiness before adoption will be significantly better placed to realize the transformation value of Generative AI while minimizing risks. Begin with a pilot, measure relentlessly, and scale deliberately.

CHAPTER 5

Building a Winning Generative AI Strategy

Introduction

Generative AI initiatives in an enterprise need a well-defined and governed strategy with a plan to harness its potential for economic growth while mitigating all its risks related to security, ethics, and workforce displacement. Without a governed framework, the potential cost of unmanaged Generative AI usage can be very high. It can also lead to serious compliance violations and profound brand damage due to AI-generated errors or biases. To avoid these pitfalls, you can't just experiment casually.

Generative AI needs a comprehensive strategy that integrates people, organizational factors, and technologies to drive business value and deliver a tangible return on investment (ROI) while mitigating risks. An effective strategy is essential for sustaining a competitive advantage and driving meaningful business transformation. It can help harness the value of Generative AI more quickly and stay ahead in the dynamic, rapidly evolving landscape.

© Brajesh De 2026
B. De, *Generative AI for Business Innovation*, https://doi.org/10.1007/979-8-8688-2682-5_5

In this chapter, we will cover the strategic aspects that must be considered at the outset to ensure the successful implementation and rollout of Generative AI within the organization. Business leaders will know how to select the right Generative AI use cases that solve real and high-value pain points. They will learn how to build the right strategy to ensure data quality and build a scalable technology infrastructure. The chapter delves into how to build guardrails for data privacy, ethics, and regulatory compliance to mitigate security risks. Strategy is incomplete without people, talent, and culture. Hence, this chapter will cover the principles and processes to build responsible, ethical, and sustainable AI solutions.

Executive Insight: *Generative AI is not just another IT tool—it's a business model enabler.*

Vision and Strategic Alignment

- **Aligning GenAI Initiatives with Business Goals**: For example, reduce the claim processing turnaround time in an insurance company by around 30% to 40% using a Generative AI-powered "Claims Co-pilot."

- **Identifying High-Value Use Cases**: For example, the "Claims Co-pilot" can speed up the claims processing steps by reading uploaded documents, extracting claim information, and creating a summary assessment report.

- **Executive Sponsorship and Cross-Functional Collaboration**: For example, the COO can declare the claim-processing efficiency improvement as a clear business mandate and approve dedicated funding. A steering committee must be established to oversee the entire program and provide governance oversight that removes all friction between cross-functional teams with clearly defined roles and responsibilities.

Some of the key questions that every CTO and CIO must ask before investing in a Generative AI initiative are

- What specific business problem are we solving, and what is the measurable value?

- What data, governance, and infrastructure capabilities are required to support an enterprise-grade AI solution?

- How will we proactively manage the ethical, security, and human risks of Generative AI to build trust with our employees, customers, and partners?

- What is our operational plan for skills and change management?

- How will we manage costs and scale our experiments?

- How does the Generative AI initiative align with the overall technology architecture and long-term roadmap?

Figure 5-1 shows the key challenges faced by CTOs and CIOs in their journey for Generative AI adoption.

Figure 5-1. *Key Challenges Faced by Executives in Generative AI Adoption*

A winning Generative AI strategy needs the following:

- **A Clear Vision**: Generative AI initiatives should be aligned with strategic business priorities to drive value.

- **Focused Implementation of the Right Use Cases**: Selecting the right use cases for Generative AI requires consideration of potential business value, feasibility, and associated risks.

- **Consistently Measure and Track Progress**: Metrics to track the performance and value of AI initiatives must be clearly defined.

Executive Takeaway: *Align Generative AI initiatives with strategic business goals with measurable outcomes. Ensure executive sponsorship and cross-functional collaboration to build ethical and secure solutions.*

Identifying and Prioritizing Use Cases

Generative AI should be deeply integrated into the overall corporate strategy and business processes to boost productivity. IT and business teams need to work closely together. Defining the right strategy must start with understanding the business goals and aligning them to the Generative AI initiatives. As shown in Figure 5-2, a dual-pronged approach, combining top-down and bottom-up initiatives, can help align corporate strategies with AI themes.

Leadership must collect feedback from the ground to understand the pain points and potential Generative AI use cases that can be implemented for productivity and efficiency improvements or to enhance customer experiences. Instead of focusing on a single AI implementation, combining several use cases can deliver significant and better outcomes and can even work to reimagine the entire value chain. A top-down approach can help pinpoint strategic domains, such as departments, business processes, or key products, that align with strategic business priorities. Merging the high-level strategy with tactical technology use cases ensures that Generative AI implementations support broader strategic objectives.

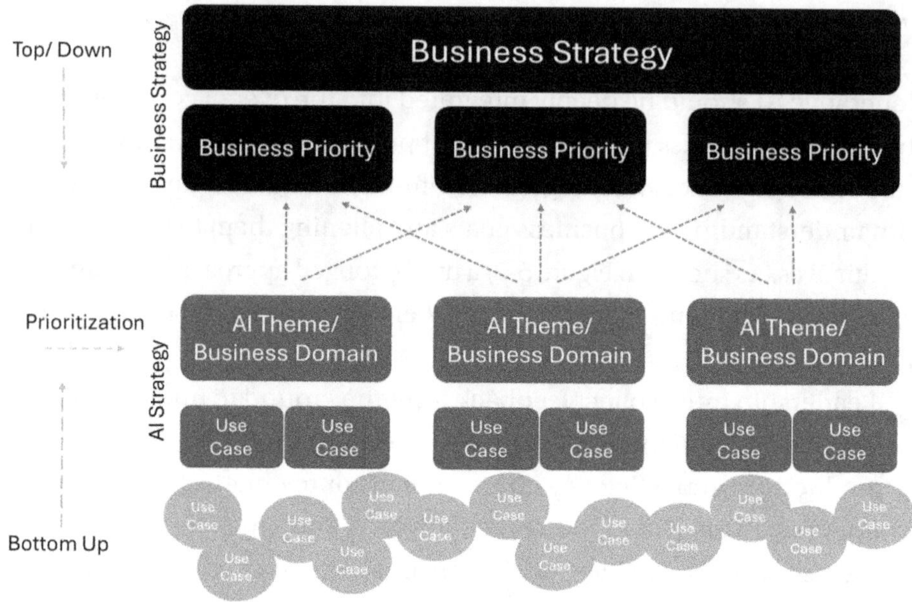

Figure 5-2. *Meeting in the Middle Approach to Identify Generative AI Use Cases*

Prioritize the Right Use Cases

Identifying the right use case for Generative AI that holds the most potential is always a challenge for business and technology stakeholders. The question that comes to mind is where to start the investment. Additionally, the organization must also assess technical feasibility, data readiness, and stakeholder buy-ins to make the final decision. One of the approaches to identify the right use case is to plot them on a three-dimensional space—**Business Value**, **Technical Feasibility,** and the **Risks.**

A simple prioritization scoring formula can be

Priority Score = Business Value x Technical Feasibility x Adjusted Risk

Each dimension can be scored on a scale of 1 to 5, with 1 being the lowest and 5 being the highest.

The following are some of the questions that can help to assess the business value or impact of the use case:

- What is the potential for revenue generation?

- What would the operational cost savings be if the use case is implemented?

- Will this have a significant positive impact on customer and employee needs?

- How does the use case support the top business objectives?

Table 5-1 provides guidance for scoring the Business Value.

Table 5-1. *Guidance for Scoring Business Values for Generative AI Use Cases*

Score	Meaning
5	Direct revenue or cost impact with improvement in customer experience
4	Significant operational efficiency improvements
3	Moderate productivity improvements
2	Nice to Have
1	Minimum measurable value

Even if the use case has a high business value, its feasibility can limit the impact. Hence, assess the technical feasibility by asking the following questions:

- Is the right quality of data available for this use case?

- Does the LLM model provide the right levels of accuracy?

- Can the available infrastructure scale to meet the future demand?

Additionally, assess operational readiness by examining talent and skill availability, change management, and the ease of integration with existing processes. Risk tolerance due to inaccurate results or misuse should also be a factor for use-case selection.

Overall, the assessment of Generative AI use cases should evaluate their business value, actionability, and feasibility.

Table 5-2 can be used as a guide to assess and score the feasibility that measures the technical and organizational readiness.

Table 5-2. *Guidance for Scoring Feasibility for Generative AI Use Case Implementation*

Score	Meaning
5	Data is ready, workflow is well understood and integration complexities are low, skills are available
4	Minor to moderate efforts needed for integration
3	Moderate data preparation and engineering efforts will be needed
2	Significant data and integration gaps to be addressed
1	Extremely complex and unclear on how to proceed

While evaluating the use cases, the associated risks should also be considered. CXOs must evaluate every Generative AI initiative against cybersecurity, financial, operational, regulatory, reputational, and vendor lock-in risks. Questions that should be asked to assess the risks are as follows:

- What are the regulatory requirements for the Generative AI use case implementations?

- What are the risks for customers using the AI application?

- What are the financial costs for model licensing, infrastructure setup, data preparation, and human oversight?

- What are the operational risks of rolling out the solution?

- What are the reputational risks if the AI decisions are biased, incorrect, or offensive?

- What are the cybersecurity risks in case of prompt injection, data leakage, or model inversion attacks?

Table 5-3 can be used to assess compliance, operational, and reputational risks.

Table 5-3. *Guidance for Scoring Risks for Generative AI Use Cases*

Score	Meaning
5	High regulatory exposure and legal and financial risks
4	Customer facing and reputational risks are possible
3	Internal manageable risks
2	Limited operational risks
1	Very Low risks

For simplicity, the Adjusted Risk Score will be calculated as follows:

Adjusted Risk = (6 − Risk)

Table 5-4 provides an example of how the different uses cases to improve the claim processing efficiency can be assessed and prioritized.

Table 5-4. *Scoring Approach to Assess and Prioritize Generative AI Use Cases*

Use Case	Value (1–5)	Feasibility (1–5)	Risk (1–5)	Adjusted Risk (6-Risk)	Final Score
Summarize claim document	5	4	2	4	80
Draft email	4	5	3	3	60
Retrieve similar claims	4	3	2	4	48
Automate claim decision	5	2	5	1	10
Assess fraud in claims	3	3	3	3	27

Final Score = Value × Feasibility × Adjusted Risk

Based on the above prioritization analysis, it is clear that the following should be the priority for the use case implementation:

1. Summarize claim document (80)

2. Draft email (60)

3. Retrieve similar claims (48)

4. Assess fraud in claims (27)

5. Automate claim decision (10)

Weighted Scoring is another approach that can be followed for prioritization of the use cases. For example, a use case to build a virtual AI agent that can answer commonly asked questions and route customers can offer high business value by reducing contact center operating costs across thousands of agents. Implementing them using Google Cloud AI offers a low-friction and highly feasible replacement, accelerating adoption and time to realize value. Hence, such a use becomes the most suitable candidate for the pilot implementation of Generative AI.

A weighted scoring framework like the one in Table 5-5 can be useful for calculating individual scores in various use cases and prioritizing them.

Table 5-5. *Weighted Scoring Framework for Use Case Prioritization*

Factor	Weight (%)	Score (1–5)	Weighted Score
Business impact	30%	4	1.2
Revenue potential	20%	5	1.0
Cost savings	20%	3	0.6
Technical feasibility	15%	4	0.6
Strategic alignment	15%	5	0.75
Total	100%		**4.15/5**

Executive Takeaway: *Prioritize AI investments using a structured framework that balances business value, technical feasibility, and risk. Focus on pilot use cases with clear ROI and ease of implementation. Begin with high-value, low-risk pilots that can demonstrate measurable ROI quickly.*

Data Strategy for Generative AI

Generative AI output is as good as the data it relies on. A high quality of diverse and well-prepared data ensures accurate, reliable, and unbiased outputs from Generative AI models. Poor data quality, on the other hand, can lead to flawed outcomes, regulatory non-compliance, financial losses, or reputational damage. On the contrary, high-quality data can reduce manual interventions and accelerate innovations.

CTOs and CIOs must focus on the following three pillars to ensure their enterprise data strategy is AI-ready:

- Data Quality

- Data Governance

- Data Accessibility

Data Quality

The key elements of data quality are

- **Accuracy and Reliability**: Accurate data is crucial for Generative AI systems to produce correct and reliable outcomes. Inaccuracies can lead to hallucination and misleading output. For example, an LLM trained on old regulatory documents can produce outputs that are non-compliant with the current regulations. If data is wrong, Generative AI will confidently produce incorrect answers.

Implement **strong data governance** and **standardize validation controls** to ensure accuracy and reliability at scale.

- **Completeness**: Completeness ensures that all critical and required data is available. Incomplete data sets can cause AI algorithms to miss essential patterns and correlations, leading to incomplete, incorrect, or biased results. For example, a medical Generative AI assistant may give unsafe treatment recommendations if patient records are incomplete. Missing regulatory data fields like KYC can lead to non-compliance. Define and enforce **mandatory data capture standards, schema validation rules** and set up continuous data completeness monitoring dashboards across all critical systems and pipelines.

- **Consistency**: Consistency ensures that data is in a uniform standard, format, and structure across systems. It helps in efficient processing and analysis of data. Inconsistent data can lead to confusion and misinterpretation, impairing the performance of the Generative AI system. For example, "U.S.", "USA", and "United States" treated differently can reduce the accuracy of semantic search in RAG systems. Set up **master data management (MDM), and centralized governance controls** to enforce uniform standards and eliminate conflicting data across systems.

- **Timeliness**: Data freshness plays a crucial role in providing contextual and up-to-date responses for time-sensitive data. Outdated data would produce outdated outputs. For example, a Generative AI-based financial assistant must rely on the most up-to-date financial information when responding to queries. Relying on outdated data may result in incorrect advice that is not relevant to the current context. Similarly, if a Generative AI system informs passengers that the flight is "On-time," while the operations team updated a delay ten mins ago, can lead to loss of customer trust and cause confusion. If data is outdated, the AI will mislead customers and break trust. Implement **real-time data pipelines, SLA-driven refresh cycles, and automated latency monitoring** with alerts to ensure data is delivered accurately and on time.

- **Validity**: Data should conform to the defined rules, formats, and business requirements. Invalid data can cause nonsensical, misleading, and risky outputs. Data validity ensures that data is meaningful, correct, and usable. Without valid data, Generative AI may generate outputs that appear accurate but are fundamentally flawed. This can lead to compliance breaches, customer dissatisfaction, or poor business decisions. For example, negative salaries in training data can produce unrealistic values, or incorrect data ranges can recommend impossible timelines. Enforce **strict data validation rules, input controls** to enforce data conforms to business and regulatory standards.

- **Uniqueness**: Uniqueness ensures that each record appears once and only once in the dataset. Data uniqueness ensures that our Generative AI is learning from a clean, non-redundant dataset. Duplicate entries can introduce bias, amplify noise, create inefficiencies, yield inconsistent results, and lead to poor customer experience. Uniqueness directly impacts trust and efficiency in AI-powered decision-making. Implement **robust deduplication rules, unique identifier standards, and Master Data Management (MDM) controls** to prevent duplicate records across systems.

- **Bias**: Data bias occurs when the dataset used to train, fine-tune, or feed a Generative AI system is not representative of the real-world population, context, or intended use case, leading to skewed or unfair outputs. Data bias is a big business risk. It can produce biased Generative AI output, potentially leading to breaches of trust, damage to brand reputation, legal and compliance risks, and ultimately, a poor customer experience. Source data from **diverse sources**, set up **bias testing frameworks, and** conduct **continuous fairness audits** to detect and mitigate systemic bias across datasets.

- **Security and Privacy**: Data security and privacy ensure that enterprise and user data used in Generative AI systems are protected at all times (at rest, in transit, and in use) from unauthorized access, misuse, or leakage. Personal and sensitive data should not be exposed, misused, or retained without consent during AI training, fine-tuning, or usage. Enterprise must protect sensitive data before, during, and after AI processing. Without strong privacy controls, Generative AI could expose an enterprise to regulatory fines and reputational damage. Implement **zero-trust access controls, end-to-end encryption, data masking, and**

continuous security monitoring to safeguard data privacy and prevent unauthorized access.

The following are some of the main challenges C-level executives need to address to ensure high data quality:

- How to collect data from various sources with the same standards?

- How to label data accurately and efficiently?

- How to address bias in data?

- How can data be securely stored to protect it from corruption?

- How to avoid errors in data as it flows from different sources?

- How to detect anomalies in data and ensure data integrity?

Here is an executive checklist for Data Quality for Generative AI:

- Is the enterprise data **accurate and reliable**?

- Is the data **current and up-to-date** for making real-time decisions?

- Have data **bias** been assessed and mitigated?

- Can data **lineage** be traced for compliance and audits?

- Is sensitive data **protected and anonymized**?

Data Governance

Strong data governance is needed for trustworthy Generative AI outputs. It ensures reliability and accuracy. It also helps ensure compliance and ethical use of AI, manage risks, and, above all, deliver business value. A Generative AI program without strong data governance is like building a skyscraper without an engineering blueprint—it may stand temporarily, but it will collapse under pressure. Effective data governance must focus on the following:

- **Data Visibility**: With data scattered across various departments and systems, it is difficult to have a centralized view of what data is available, who owns it, and where it is stored. This often results in extra efforts to find and prepare the data for Generative AI needs. Create a centralized data catalog for all data from on-premises databases, data lakes, and SaaS applications as the first step toward data governance initiatives.

- **Access Control**: With siloed data repositories, managing access permissions to different data sources becomes complex. This creates a lot of challenges in accessing vital data and analytics owned by other departments. Implement role-based access control with 48 hour approval SLAs.

- **Ownership**: Effective data governance for Generative AI **needs a significant cultural shift**. Employees must see data control or ownership as a strategic benefit rather than a necessary burden. CTOs and CDOs must address this people challenge, which often blocks success. **A self-orchestrating governance approach** can enhance decision-making, boost efficiency, and foster growth—but only if all departments buy in and take ownership of the process. Being a data steward means employees are responsible for properly managing, securing, and sharing data, rather than hoarding it. Name one data steward per critical domain with quarterly accountability.

- **Quality Assurance**: Garbage in, garbage out. Generative AI can only give reliable results if the data feeding it from different systems and platforms is of high quality. Low data quality increases the likelihood of errors and inconsistencies. **Automated data quality controls** must be implemented to ensure data integrity. **Data quality stewards must be appointed** for all domains to set standards, implement quality checks, and monitor data health. Weekly dashboards and **monthly audits must be performed** to ensure compliance with regulatory standards and reduce vulnerability to legal risks and reputational damage.

Key Takeaway: *Data Governance must be a C-suite priority. For effective data governance, an organization must create a Chief Data Officer (CDO) role reporting to the CEO and establish a cross-functional Data Governance Council.*

Data Accessibility

Data is a strategic enterprise asset and should not be locked in silos. Data accessibility refers to the ease with which people can find, retrieve, understand, and use data to gain insights and make decisions. It involves making data available and usable for authorized users, including employees, partners, or citizens, without unnecessary technical barriers. This process includes making data comprehensible through standardized formats, efficient retrieval systems, and user-friendly tools, ultimately enabling informed decisions, and increased productivity.

Data should be secure by design—balancing innovation (broad access) with risk management (controlled access). Data should be categorized into public, internal, confidential, and restricted, and access can be enabled via APIs and data catalog secured using role-based access control. Generative AI systems should be able to discover and request access to curated datasets.

Executive Takeaway: *Ensure high data quality and establish strong data governance practices to build fair, secure, and explainable Generative AI solutions.*

Model Selection and Adaptation Strategy

LLMs are the backbone of all Generative AI applications due to their reasoning abilities and the capability to generate human-like language. Trained on different types of data, different LLMs have different competencies and strengths. Some models are best at text tasks like summarization or translation, others at code generation, and others at data analysis. They also differ in their performance characteristics, computing needs, and costs. Selecting the right LLM model for a given use case is often the first step for any Generative AI implementations.

A structured, methodical approach is needed for selecting the right LLM. Enterprises have a range of base models and their variations to choose from. Organizations must make the right choice based on multiple factors to ensure success for their Generative AI implementation. For example, simple text summarization can be done by a low cost and fast model, while code generation would need a higher accuracy model.

Selecting the right LLM can be overwhelming at times, as the right LLM must offer the best balance of performance, accuracy, cost-effectiveness, scalability, data security, and strategic fit with the organization's needs. It involves first selecting the appropriate model category and then choosing a suitable LLM within that category.

Model Category Choice: Proprietary vs. Open Source vs. Domain-Specific vs. SLMs

There are different categories of Generative AI models as follows:

- **Proprietary Models** are developed and maintained by a specific company. Access to these models is restricted and is generally granted through APIs or licensing agreements. They provide stable enterprise-ready performance and are backed by rigorous testing and vendor support. However, they lack flexibility and can be potentially costly as usage scales. Examples include OpenAI, Claude, and Google Gemini. Adopt proprietary LLMs when you need **enterprise-grade security, regulatory compliance, faster time-to-value, strong vendor support, and have minimal customization needs.**

- **Open-source Models** refer to pre-trained models whose source code and trained model weights are publicly available for anyone to use, modify, and distribute. Because the source code is open source, it provides transparency on how the model works and accelerates innovation and development of new AI applications. Examples of open-source models include LLaMA, Falcon, and Mistral. Select open source LLMs when you need **greater customization, cost control at scale, data residency controls, and want to avoid vendor lock-in.**

- **Domain-specific Models** are language models that are designed, trained, and optimized for a specific industry. They are trained on specialized and curated datasets from the target industry, such as finance, healthcare, legal, engineering, etc. This gives the model a deep understanding of the concepts and terminology of the specific industry. They may use algorithms specifically designed to address challenges in that domain. Examples include BloombergGPT, BioGPT, and ChatLaw. Select domain-specific models when you need **high accuracy in specialized workflows, reduced hallucination risk, and optimized performance for industry-specific** data and terminology.

- **Small Language Models** are language models that are trained on a smaller dataset to perform and excel in specific tasks like code generation or financial modeling. They have fewer parameters than LLMs, typically ranging from a few million to a few billion, compared to LLMs with hundreds of billions or even trillions. SLMs need less computational power, memory, and energy, making them suitable for resource-constrained environments like mobile devices. They are generally much faster, cost-effective for training and inference. Examples of SLM include codeGen from Salesforce, GPT-4o mini, Microsoft Phi series, and Google Gemma. Select small language models when you need **low-latency, cost-efficiency, and edge-deployable AI** for focused tasks with **tighter data control and predictable performance**.

Models Evaluation Criteria

With several models in each category, the next step for any organization is to select the best model that meets its requirements. The following are some of the main factors to be considered while selecting the right LLM model:

- **Functional Capabilities**

  - **Multi-Modal Support**: Can the model handle complex documents and varied inputs (text, images, etc.)?

  - **Customization/Fine-Tuning**: Does the model support fine-tuning or customization for domain-specific needs?

  - **Enterprise Workflow**: Does the model handle complex enterprise workflows reliably

- **Performance and Quality**

  - **Accuracy**: Are the responses from the model correct and reliable?

  - **Relevance**: Does the model provide contextually relevant responses?

  - **Clarity**: Is the model response/output easy to understand?

  - **Latency and Speed**: Does the model meet the latency and throughput requirements?

  - **Benchmark**: How does the LLM perform against industry standards?

- **Responsible AI and Ethical Considerations**

  - **Fairness**: Does the model provide impartial output with no favoritism?

  - **Transparency**: Is there clarity about the training data, input data handling, and model architecture?

  - **Explainability**: Can the model's decision-making process be understood to address potential ethical concerns?

- **Cost and Economic Viability**

  - **Cost Efficiency**: What is the cost for training, inference, fine-tuning, and ongoing infrastructure maintenance?

  - **Computational Resource**: What are the infrastructure requirements for inference and training?

  - **Pricing Model**: What are the subscription and usage costs and licensing terms?

- **Security and Compliance**

  - **Data Privacy and Compliance**: Does the model handle sensitive data and meet regulations?

  - **Access Control**: What mechanisms are used to enforce authentication and authorization for model access?

  - **Vulnerability to Attack**: How is the model protected from prompt injection, adversarial attacks, or data poisoning?

- **Integration and Deployment**

  - **Ease of Integration**: Does the model seamlessly integrate with existing systems and APIs?

  - **Deployment** : Does the model support flexible deployment options for on-premises, cloud, or hybrid setup?

- **Vendor Maturity and Ecosystem**

  - **Provider Reputation**: What is the reputation of the LLM provider for quality, reliability, and ethical practices?

  - **Documentation Quality**: Are there clear and thorough documentation and guidance for using the model?

  - **Support**: What customer and community support are available to resolve any issues with using the LLM?

LLM Benchmarks

LLM benchmarks provide standardized evaluations of capabilities such as reasoning, logical thinking, language understanding, translation, math, and coding.

The following are some of the most common LLM benchmarks that can be looked at to assess the raw capabilities for evaluation:

- Knowledge: MMLU and TriviaQA

- Translation: Chatbot Arena, BLEU

- Reasoning: HellaSwag and WinoGrande

- Comprehension: DROP and Race-h

- Coding: HumanEval and MBPP

- Math: GSM8k and MATH

However, strong benchmark scores do not guarantee real-world performance, like user experience that includes latency, context management, and the ability to adapt to user-specific needs. They are "screening tests" and must be paired with a task-specific evaluation that measures latency, context window performance, and failure modes in your own data.

Model Adaptation Framework: Training, Fine-Tuning, and Retrieval Augmented Generation, Prompt Engineering

LLM Models may need to be adapted to customize them for a specific domain, tasks, or user needs to enhance their relevance, accuracy, and usefulness. Training, Prompt Engineering, Fine-Tuning, and Retrieval Augmented Generation (RAG) are different ways to customize LLMs. Customization and adaptation of LLMs improve overall performance and enable personalized user experience. Training is the foundational step to building a base model. Fine-tuning further trains the base model on specialized datasets to improve domain-specific skills and style. Retrieval Augmented Generation provides dynamic access to external data, allowing for up-to-date, contextually rich, and factual information without altering the model itself. Figure 5-3 illustrates the cost and computational complexity of model adaptation.

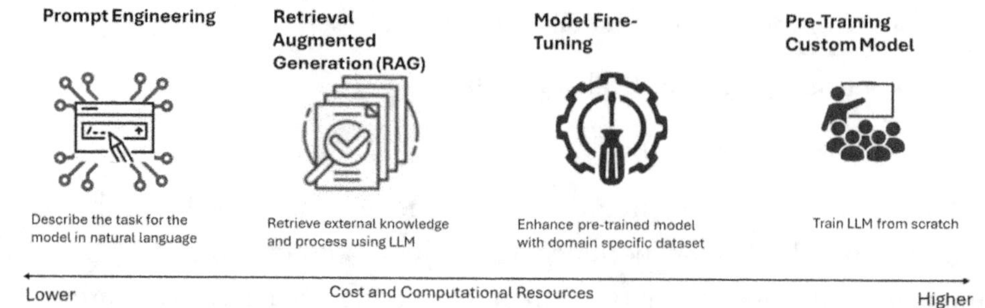

Figure 5-3. *Cost and Computational Complexity for Model Adaptation*

Training AI Models for Complex or Unique Business Problems

Foundational models are large, pretrained models trained on massive unlabeled datasets across multiple domains. These models are capable of solving a wide range of downstream tasks with little or no additional training. However, some use cases may require a custom model that needs to be trained specifically for a defined task using your own data. This provides greater control, improved data privacy, lower cost, and better performance on narrow tasks.

However, training AI models would require managing and processing vast amounts of diverse and generalized datasets to establish their core knowledge and capabilities. The data volume can be in petabytes. Training the AI models requires a lot of computational power. In particular, training large language models (LLMs), image generators, and multi-model systems requires highly scalable compute resources that can efficiently handle high volumes of data, making it expensive. For enterprises, the ROI of training from scratch is rarely justified, and hence should be carefully thought through.

Some of the key challenges that the data architecture for pre-training Generative AI models must address are

- Efficient and cost-effective collection of labeled data from diverse sources

- Ensure high data quality across heterogeneous data sources

- Providing scalable compute resources and efficient storage infrastructure

- Building efficient retrieval systems for large-scale datasets

Other key factors to consider include

- Data versioning for traceability and reproducibility

- Privacy protection for broad and sensitive datasets

- Sustainable practices to manage energy usage and long-term storage cost

- In-house expertise in machine learning for training models

Prompt Engineering for Adaptability, and Bespoke AI-driven Outcomes

Prompt Engineering is a technique of crafting the input prompt to guide the Large Language Model to provide the desired response. It is a way to influence and communicate with AI so that it understands the intent and produces accurate, relevant, and high-quality results.

Prompt engineering has the following advantages:

- **Quality:** Improved AI performance with more accurate and relevant responses

- **Consistency**: Greater control and predictability for consistent responses

- **Safety**: Bias Mitigation, as clear prompting can reduce bias and inappropriate responses

Prompt Engineering is low-cost, fast, and can be easily implemented with minimal expertise and integration needs.

Retrieval Augmented Generation (RAG) to Power Real-Time Response

Retrieval Augmented Generation (RAG) enables real-world applications to deliver accurate, context-aware, and up-to-date responses. It combines the power of LLM with enterprise-specific data sources. Figure 5-4 demonstrates the RAG architecture that first retrieves the relevant information from the enterprise data sources and then augments it with LLM's knowledge to produce the final response. This approach helps eliminate issues that can arise from outdated information in LLMs.

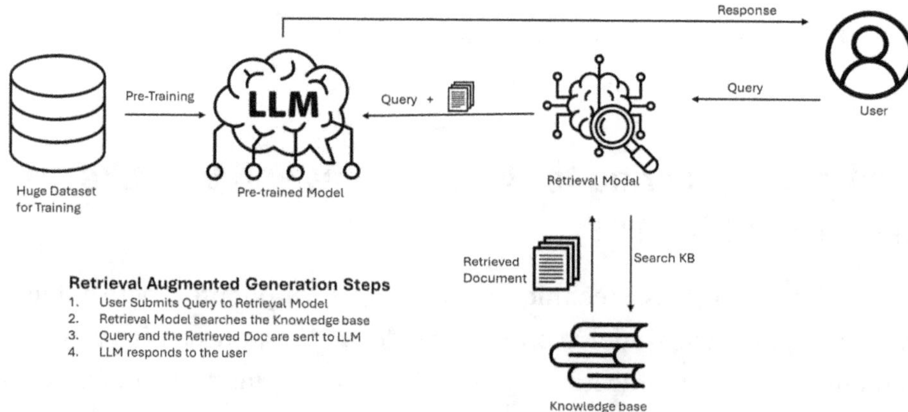

Figure 5-4. *Retrieval Augmented Generation*

Below are some real-world use cases, where RAG systems can be used:

- **Customer Support**: Chatbots built using RAG can provide accurate and timely answers to customer queries.

- **Legal Research**: Lawyers can use RAG to quickly find relevant information from legal documents, statutes, and case law.

- **Research Assistant**: Researchers can stay up-to-date and quickly find relevant information from a database of research papers using RAG in a concise form.

- **Digital Medical Assistant**: Doctors and healthcare professionals can access the latest research and clinical guidelines to support diagnosis and treatment decisions using RAG.

- **Enterprise Knowledge Base**: A RAG system can query internal documents, knowledge bases, and policy information to provide the required information for employees more easily.

Supervised Fine-Tuning Models for Domain Adaptation and Specific Needs

Once foundational models are trained, supervised fine-tuning tailors a pre-trained general-purpose LLM for a specific domain or task. This transforms AI from a cost center (an API expense) to a strategic capability and a core differentiator. Validated examples are used to train the model to learn and answer better to prompts. This requires sourcing smaller, high-quality domain or task-specific labeled datasets that are

- Sourced from internal systems, public datasets, or trusted third-party providers

- Filtered to remove irrelevant data

- Anonymized to meet data privacy and compliance requirements

- Free from bias in the training dataset

- Formatted appropriately for the chosen model and platform

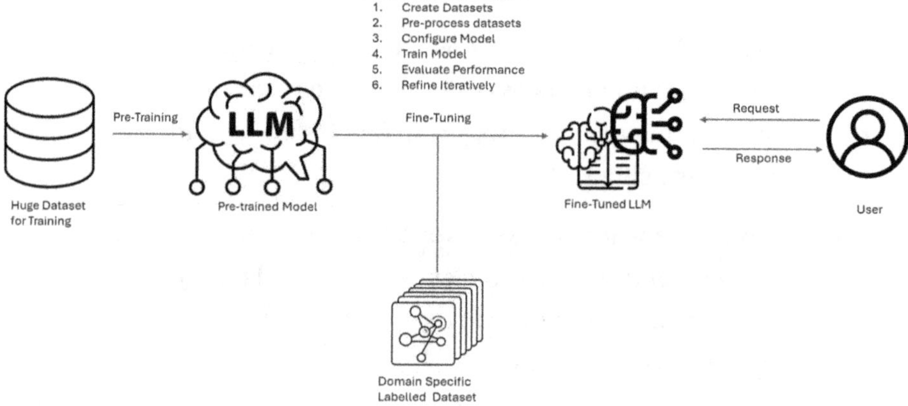

Figure 5-5. *Fine-Tuning*

Figure 5-4 illustrates how supervised fine-tuning adjusts the weights of a pre-trained model by teaching it task-specific nuances from labeled data, drastically improving its performance on a specific task. Supervised fine-tuning of models provides the following advantages:

- Better answers, matching your business guidelines

- Competitive advantage with the company's specific knowledge

- Improved output quality, tone, and brand safety

- Improved operational efficiency and optimized cost in the long run

Fine-tuning a model becomes the preferred choice in the following scenarios:

- The application needs specialized domain knowledge. It is useful for drafting legal contracts, creating medical reports, or generating code in a specific language.

- Improve the base model's weakness.

- The application has deployment constraints that prevent external data access.

However, the downside of supervised fine-tuning is that it requires powerful GPUs for training, along with specialized skills and talent in LLMOps and LLM training techniques such as LoRA or QLoRA.

Which LLM Adaptation Approach to Choose

The adaptation approach to choose from would depend on the business problem to be solved and the expected ROI (efficiency, customer experience, cost reduction, innovation). Table 5-6 provides a small framework to help you choose the right one for your business needs.

Table 5-6. *LLM Adaptation Framework*

Evaluation criteria	Prompt engineering	Retrieval augmented generation (RAG)	Supervised fine-tuning
When to use	Get the desired response from the pre-trained LLM.	Secure enterprise knowledge is required to perform the task.	Specialized domain knowledge is required to perform a specific task.
Data needs	No special data is needed. Only examples and input prompts.	Access to curated and indexed enterprise data would be required for the LLM to respond.	Domain-specific labeled data would be needed to fine-tune the LLM.
Data governance focus	Avoid PIIs in input prompts. Have guardrails in place and monitor output data for responsible usage.	Data should be fresh with trackable lineage. Access to data should be controlled.	IP and ownership of the training data should be protected. Training data should be versioned, auditable, and free from bias.
Infrastructure needs	No custom infrastructure needed beyond API gateway, authentication, and monitoring. Integration complexity is low.	Needs a vector database and ETL document pipelines for unstructured data to flow to the vector store.	Higher infra complexity—requires ML pipelines, GPU/TPU resources. Needs model registry, versioning, and deployment pipelines.

Integration complexity	Low	Medium	High
Skills needs	Low barrier to entry, minimal ML talent required. Business users can iteratively design effective prompts.	Needs data engineers, vector database engineers, LLM orchestrators, Knowledge managers, and MLOps engineers to keep the knowledge base fresh and accurate.	Needs ML engineers, data scientists, MLOps engineers, QA specialists, and legal experts to deliver high-value domain expertise. Retaining talent can be costly.
Talent availability	Easy to upskill existing staff.	Moderate, available in cloud/data orgs.	Harder to hire/retain.
Cost	Quick win with low cost.	Balanced cost and scalable.	Fine-tuning the LLMs can be costly.

Executive Takeaway: *Evaluate and select the right model based on functional capability, performance, ethical considerations, cost, and security. Adopt a model adaptation strategy based on data availability, skills, and ROI expectations. Start with prompting; move to RAG or fine-tuning only if prompting cannot meet accuracy, compliance, or cost targets.*

Generative AI Technology Strategy

Creating the right foundational technical architecture is needed for scaling Generative AI solutions responsibly, securely, and cost-effectively. Only then can it deliver real business values. Building a scalable Generative AI architecture is more than just using the right LLM model. Figure 5-6 shows the Generative AI technology stack. It needs a layered approach built on a robust infrastructure, a flexible platform to hold and run the LLMs, high-quality data, a data pipeline to train and enhance the models, tools, and techniques used to coordinate and execute an autonomous workflow, and an interface for users to access the LLMs.

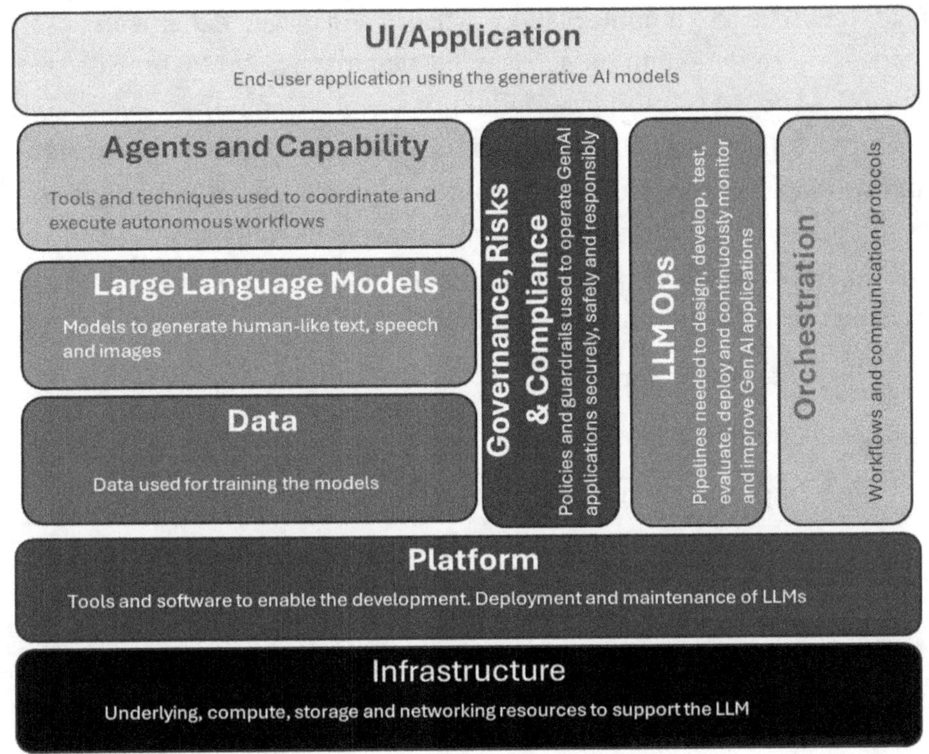

Figure 5-6. *Generative AI Technology Stack*

Infrastructure Layer

Generative AI applications need to manage large datasets and complex computations. Hence, it needs a robust infrastructure that provides high-volume data storage, parallel processing, and high-performance AI computational power to provide a low-latency response in real time. The infrastructure resources can be physical or cloud-based. The physical resources can be on-premises with self-managed LLMs deployed on dedicated hardware like NVIDIA DGX or GPU Clusters. In a cloud-based deployment, the LLMs can be self-managed or hosted APIs. Cloud hyperscalers, such as Azure, AWS, or Google Cloud, provide options for organizations to deploy

189

LLMs on a scalable compute platform, such as a managed Kubernetes service or serverless compute services. Popular managed LLMs provided by OpenAI (ChatGPT), Anthropic (Claude), or Google (Gemini) are available via managed APIs. Hosting LLMs on-premise or on a cloud-hosted Kubernetes infrastructure provides more control but requires high maintenance efforts. On the other hand, managed LLMs may be easy to use but introduce vendor lock-in. The following are some of the options that one should consider for choosing the right deployment model for LLMs:

- Integration complexity

- Compliance and regulations

- Implementation time

- Cost

- Availability and resilience

Table 5-7 can help you choose the right LLM deployment model that is best suited for your organization.

Table 5-7. *Framework to Select the Right LLM Deployment Model*

Factors	On-premises (self-managed LLM)	Cloud-based (self-managed LLM)	Cloud-based (hosted API LLM)
Integration complexity	High integration complexity. Organizations need to set up their own infrastructure with GPUs for model serving and inference.	Moderate complexity; Cloud providers provide compute resources with high capacity and GPUs. Organizations will still have to manage the complexity of model serving and inference.	Low complexity; LLMs are exposed as API endpoints, enabling faster integration with minimal effort.
Compliance and regulations	Ideal for highly regulated industries like finance or healthcare that require complete control.	Cloud providers provide compliance support. Organizations have to abide by the shared responsibility model for data security.	Compliance depends on provider certifications. Organizations have to ensure that data protection policies are met.
Implementation time	Longer implementation time as Organizations have to produce and set up their own hardware.	Moderate. Organizations have to configure the cloud infrastructure for deploying the LLMs.	Fastest. LLM inference services are available as ready-to-use APIs with minimal configurations.

(continued)

Table 5-7. (*continued*)

Factors	On-premises (self-managed LLM)	Cloud-based (self-managed LLM)	Cloud-based (hosted API LLM)
Cost	Predictable cost but needs large up-front investments.	Cost fluctuates based on storage and compute requirements.	Subscription-based pricing can simplify budgeting. However, costs can increase rapidly with high usage.
Availability and resilience	Depends on the high-availability cluster setup of the infrastructure and the disaster recovery processes.	Cloud providers offer high availability, but organizations must set up resources for failover.	Redundancy is built in and managed by the provider.

Platform Layer

The platform layer consists of the software and tools used to develop, train, and deploy LLM models. These include programming languages like Python, libraries such as PyTorch and TensorFlow, and specialized platforms like Hugging Face Transformers, LangChain, and LlamaIndex. For deployment, tools like MLflow, TensorFlow Serving, and cloud-based solutions like AWS SageMaker and Azure Machine Learning are popular.

Executive Takeaway: *Build a scalable, secure, and flexible tech architecture that supports rapid experimentation and future scaling.*

Governance, Risk, and Compliance

- AI Risk Management Framework (NIST AI RMF) **4**

- Ethical AI principles and bias mitigation

- Regulatory landscape and enterprise GRC practices

- Human-in-the-loop and oversight mechanisms

Generative AI should be designed and implemented for a safe and productive user experience. It should be built in accordance with ethical principles and comply with local laws and regulations. Figure 5-7 shows the core principles for Generative AI.

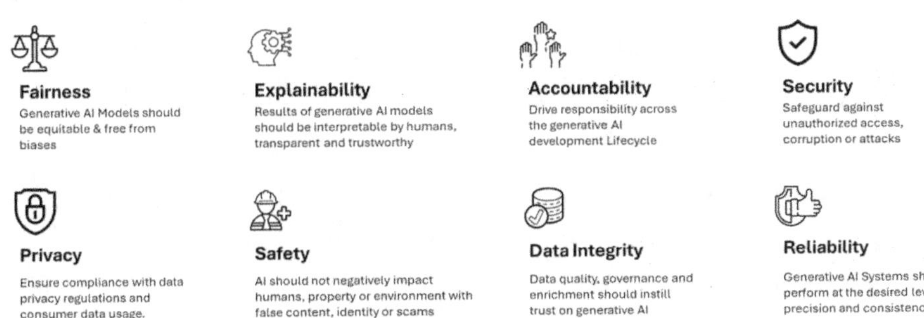

Figure 5-7. *Core Principles for Building Generative AI Solutions*

It is essential to foster a culture that promotes the responsible and ethical use of Generative AI. Below are some of the guidelines to promote responsible AI Governance, set accountability, and promote ethical use of Generative AI.

Transparency and User Trust

Foster clarity and trust with both users and the public.

- **Disclose AI Involvement**: Clearly identify and call out the origin of AI-generated content when shared publicly to foster trust.

- **Explain Outputs**: Ensure the output of Gen AI systems is explainable to users for them to understand the influencing factors.

- **Prevent Misuse**: Ensure clear communication regarding AI-generated content to prevent confusion and deception.

- **Foster Digital Literacy**: Provide training and empower users with guidelines and best practices to assess AI-generated content critically.

Security, Privacy, and Compliance

Protect sensitive data and ensure robust security measures are in place.

- **Safeguard Sensitive Data**: Implement data encryption, access control, and security audits to safeguard against potential breaches.

- **Stay Compliant**: Regularly align with regulatory requirements (GDPR, HIPAA, AI Act, etc.) and industry standards.

- **Anonymize by Design**: Anonymize, declassify, and eliminate any sensitive data used for training models.

- **Update Regularly**: Schedule regular updates to improve functionality, security patches to protect against new threats, and stay compliant.

Operational Excellence and Human Oversight

Maintain high-quality, accurate outputs through human-in-the-loop processes.

- **Embed Human-in-the-Loop**: Ensure human oversight and review to maintain the accuracy, quality, and relevance of AI-generated content.

- **Maintain Editorial Control**: Validate content for accuracy, quality, and contextual relevance prior to release.

- **Enable Feedback Loops**: Capture learnings from users, operators, and auditors to refine AI systems.

- **Evolve Responsibly**: Treat Generative AI as living assets requiring life cycle management and schedule regular updates to improve the functionality and address issues.

- **Enable Feedback Loops**: Validate and monitor Generative AI results and evolve continuously.

Executive Takeaway: *Implement a responsible AI framework with clear accountability and explainability to build safe and secure Generative AI solutions that users can trust. Embed human-in-the-loop processes for critical decisions.*

Measure Performance and KPIs of Generative AI

Evaluating the performance of Generative AI is important to ensure that it is delivering its promised outcome. Key Performance Indicators (KPIs) are crucial for objectively assessing performance and demonstrating tangible returns. The following are some of the **technical metrics** that can be used to measure the progress of Generative AI initiatives:

- **Model Quality Metrics** help to evaluate and monitor the accuracy, factuality, reliability, and security of the Generative AI models. High model quality reduces rework, errors, or compliance breaches.

- **System Metrics** monitor the health and performance of the Generative AI platform and infrastructure. It uncovers bottlenecks and areas for performance optimization, helping to ensure that Generative AI systems run efficiently, reliably, and at scale. It is also a measure of improved customer satisfaction due to faster service.

- **Adoption Metrics** offer insights into adoption rates, usage frequency, and other forms of qualitative user feedback. It is helpful to understand how users are utilizing Generative AI tools to improve their productivity.

- **Operational Metrics** measures whether Generative AI features and capabilities are delivering downstream returns. It helps to understand the impact on business processes.

Measuring KPIs across these four areas provides a holistic value of Generative AI, covering the technical impact of the investments made. However, to understand the business impact of Generative AI, a C-suite executive must review the following **business metrics**:

- **Revenue Growth**: The percentage increase in upsell/cross-sell of conversions or reduction in time to market for new products due to Generative AI initiatives or increase in customer lifetime value (CLV) are some of the parameters to review the revenue growth.

- **Cost and Efficiency Gains**: Reduction in operational or infrastructure cost or the process cycle-time reduction can be reviewed to measure the cost and efficiency gains due to Generative AI.

- **Customer and Employee Experience:** The net promoter score, or improvement in first contact resolution in customer service, or the percentage improvement in employee productivity can provide insights into customer and employee experience due to Generative AI.

- **Risk, Compliance, and Trust:** Reduction in compliance breaches, errors, or hallucination in AI output, or the auditability and explainability scope of Generative AI output can help to assess the risk, compliance, and trust improvement with AI.

Comparing these KPIs against baseline values can help gauge the effectiveness of investments in Generative AI. Establishing consistent practices and methods for measuring these metrics is a challenging task. The best practice is to develop this measurement plan during the design phase of the use case.

Executive Takeaway: *Stating AI goals clearly is key to encouraging and enabling organization-wide fluency and adoption of AI. Define technical KPIs that are linked to business outcomes.*

Operationalization and Scaling Generative AI (GenAIOps and FinOps)

Many organizations build AI models that work well in limited environments but struggle to scale. Scaling a Generative AI solution from pilot or proof-of-concept to production in a sustainable manner requires further consideration. CTOs and CIOs must consider how to manage increased complexity with enhanced speed, accuracy, and quality. They must manage the operational challenges, control the spiraling costs, and implement robust governance measures to scale the solution efficiently and safely. GenAIOps addresses the operational challenges of scaling AI solutions. A robust FinOps practice is essential to track the usage of cloud resources and optimize their Generative AI spending. Continuous evaluation and multi-layered guardrails are fundamental for the responsible and effective deployment of Generative AI solutions.

GenAIOps and Observability

GenAIOps is an extension of MLOps and LLMOps to handle the special needs of building Generative AI applications. Generative AI models have the following unique challenges that make the traditional MLOps practices insufficient:

- Training Generative AI models with billions of parameters requires high compute resources and special infrastructure.

- Models should be safeguarded from producing harmful content.

- Non-deterministic outputs make testing and validation complicated and challenging.

- The Generative AI application stack includes additional components, such as prompts, vector databases used for RAG, guardrails, and evaluation frameworks.

A comprehensive GenAIOps framework should address the following to design, deploy, monitor, and scale applications effectively in production:

- **Model Life Cycle Management**

  - **Model Versioning**: Track different versions of Large Language Models and Fine-Tuned models for reproducibility and traceability

- **Experimentation**: Log hyperparameters, datasets, metrics, and outcomes of training/fine-tuning experiments using tools like MLFlow or Weights & Biases

- **Model Registry**: Centralized storage for approved models, with metadata (version, performance benchmarks, lineage)

- **Prompts and Context Management**

  - **Prompt Library**: A catalog of versioned and reusable prompt templates for different use cases

  - **Prompt Engineering**: Design and refine prompts for Generative AI models that produce the desired outputs and maximize performance.

  - **Prompt Versioning**: Manage, track, and control changes to prompt over time.

- **Model Service and Inference**

  - **Model Evaluation**: Evaluate the performance of the model based on the performance, latency, quality of response, and feedback received.

  - **Inference API**: Expose scalable API endpoints for model inference as REST/gPRC.

  - **Caching Layer**: Cache frequent queries and responses to reduce latency and minimize redundant compute cycles. Set cache TTL and eligibility rules.

- **Autoscaling Infrastructure**: Leverage Kubernetes, serverless frameworks, or managed GPU services to auto-scale inference workloads based on demand. Set autoscaling triggers based on QPS or GPU utilization.

- **Automation and CI/CD**

  - **Model Deployment Pipelines**: Automate packaging, testing, and rollout of fine-tuned models or prompt configurations

  - **Data Pipeline Automation**: Automate data ingestion, cleaning, vector indexing, and storage of datasets

  - **Alerts**: Auto-trigger alerts in case of any deviation or anomalies like output drift or high latency

- **Monitoring, Logging, and Observability**

  - **Model Performance**: Measure and track the latency, usage patterns, token usage counts, error rates, and response quality of models.

  - **Cost Observability**: Measure the cost per request and workflow and track the token consumption trends.

  - **Drift Detection**: Detect data drift and degradation of model output quality over time.

  - **Responsible AI Monitoring**: Manage regulatory and reputational risks by monitoring bias detection metrics, toxicity and harmful output scoring, and human override logs.

- **Audit Logs**: Store detailed logs of model input/output, inference events, and user interactions for compliance and debugging.

- **Dashboards**: Create dashboards to monitor model health, usage stats, drift indicators, and cost analytics in real time.

- **Feedback Loop and Continuous Improvement**

  - **User Feedback Integration**: Capture user feedback, ratings, corrections, or logs for supervised retraining.

  - **Active Learning Pipelines**: Prioritize samples for retraining based on uncertainty or user interactions.

FinOps for Cost Management

To keep the cost of Generative AI initiatives under control, it is important to track the consumption of significant cloud resources. CTOs and CIOs should have detailed visibility into usage and costs. They should do a detailed cost analysis and understand the expenditure categories, identify cost drivers, and plan for future growth. Selecting a cost-efficient model for inference can significantly reduce cost. They should also track GPU/CPU utilization, as well as the cost per inference request. A cost analysis is required to make informed decisions about model deployment and the optimization of underutilized resources. Cost-effective compute resources, such as spot and reserved instances, can also help optimize costs. The system should also be architected to reduce per-request inference costs by caching frequently queried results and utilizing batch processing. Regular reviews of performance vs. cost of models and making necessary adjustments can also optimize costs. Leveraging open-source pre-trained models and fine-tuning models over full retraining can help to optimize compute costs for model training. Using a smaller distilled model over a large full-blown model for a specific use case may prove to be more cost-effective. Overall, a CTO must focus on the following for a solid FinOps strategy for Generative AI:

- Visibility into usage and costs

- Cost-efficient model selection and serving

- Smart infrastructure scaling

- Governance for accountability

- Data life cycle cost control

- Forecasting future needs

The following are some of the other governance aspects that should be part of the CTO's strategy to scale Generative AI solutions:

- **Access Control**: Role-based access control (RBAC) should be implemented for managing LLM fine-tuning, deployment, and inference APIs.

- **Audit Logs**: Detailed audit logs should be maintained to track all model-related operations, like who trained the model, when, and using what data.

- **Data Privacy Compliance**: All data used for training, fine-tuning, or prompt enrichment must follow regulations.

Executive Takeaway: *Implement GenAIOps practices for model versioning, prompt management, inference autoscaling, and drift detection. Employ FinOps to track cloud resource usage, optimize costs, and ensure cost-effective model deployment.*

Sustainability Considerations

Generative AI initiatives can have a negative impact on sustainability if not controlled. Senior leadership should not only evaluate the business value, consider implementation complexity and costs but also scrutinize environmental impacts such as GHG emissions, electricity usage, and water consumption. The computational power required to train Generative AI models with billions of parameters can demand a significant amount of electricity. The energy consumed by computing hardware for responding to a consumer query is significantly higher than a simple web search. This can lead to increased carbon dioxide emissions and pressure on the electric grid. Additionally, a substantial amount of water is required to cool the hardware used for training, deploying, and fine-tuning Generative AI models. Hence, the focus should be on using energy-efficient hardware for deploying Generative AI applications.

Executive Takeaway: *Align sustainability with ROI and long-term strategy.*

Executive Decision Framework

Table 5-8. *Executive Decision Framework for Generative AI Strategy*

Strategic Area	Executive Question
Vision and Strategy	What is the main purpose of investing in AI?
Use Cases	Which opportunities deliver the highest value?
Data	Is our data AI-ready?
Models	Which model strategy best fits our needs?
Technology	Can our architecture scale?
Governance	Are risks adequately managed?
Operations	Can we sustain and optimize AI at scale?

Conclusion

A winning Generative AI strategy is not about chasing the latest trends or adopting technology for technology's sake. It is about making informed, strategic decisions that align with long-term business goals and deliver measurable value. Organizations must integrate Generative AI ethically and responsibly into their business operations, focusing on driving innovation, accelerating value creation, and enabling new business models.

Successful implementation starts with defining clear, outcome-driven objectives, supported by a robust data strategy and a culture that embraces data-driven decision-making. Tech leaders must act with urgency—delays in shaping the strategy will only widen the gap between market opportunities and competitive readiness. A platform-centric approach is essential to establish a Generative AI Center of Excellence, ensuring centralized governance, best practices, and scalable experimentation.

Equally critical is building a resilient, scalable enterprise technology and data platform that supports multiple AI use cases while ensuring security, compliance, and performance at scale. Alongside technology, organizations must invest in people—developing skills in AI literacy, responsible AI use, prompt engineering, and cross-functional collaboration. A people-centric culture that promotes transparency, accountability, and trust is the foundation for sustainable success.

CxO leadership must recognize that the Generative AI journey is a marathon, not a sprint. Realistic time horizons, continuous learning, and iterative scaling are key to navigating this complex transformation. By taking a structured, disciplined approach—balancing bold innovation with responsible governance—leaders can position their organizations to thrive in the age of intelligent automation and Generative AI-driven business models. Start with a 6-to-8-week pilot implementation with defined KPIs to evaluate feasibility, identify risks, and measure adoption. Establish a holistic framework, lay the foundations with the right foundational models, and scale the pilot to production.

CHAPTER 6

Generative AI Architecture and Infrastructure: Building the Foundation

Introduction

Generative AI is reshaping how enterprises create value by enhancing customer experiences, accelerating decision-making, and enabling new product innovations. Many enterprises have invested in building pilots to demonstrate the technical promise of Generative AI. While these pilots have succeeded technically, most have failed to reach production. An MIT study found that over 90% of AI pilots fail to reach production, often due to fragmented non-standardized infrastructure, limited scalability, lack of robust security, escalating costs, and an inadequate governance framework.

© Brajesh De 2026

B. De, *Generative AI for Business Innovation*, https://doi.org/10.1007/979-8-8688-2682-5_6

For technology leaders, this is not just a model challenge—it's a **strategic architecture problem**. Scaling Generative AI applications requires thoughtful considerations to build a resilient foundation that integrates data governance, security controls, cost management, and observability from day one. Without this, even technically successful pilots struggle to deliver sustained business impact.

With Generative AI, enterprises today do not need to build a new model for every use case. Instead, they can easily adapt and orchestrate existing Generative AI models for their specific use case. However, building applications powered by Generative AI presents unique design challenges-non-deterministic outputs, variability under identical prompts, and evolving model behavior. Hence, there is a need to provide guidance on how to design and build Generative AI applications with strong guardrails that ensure a safe, secure, and reliable deployment at scale.

This chapter will cover the architecture patterns and design principles for building responsible Generative AI solutions. It will also introduce the Generative AI architecture stack and cover the architectural considerations for infrastructure, platform, model operations, and security to build a scalable and resilient application using Generative AI.

Design Principles for Building Generative AI Applications

Generative AI is probabilistic in nature. Hence, its outputs can vary for the same user inputs and can occasionally be incorrect. This variability demands a fundamentally different approach to designing user interactions with Generative AI systems. This approach must actively manage unpredictability, set expectations, and mitigate risks.

Today, humans interact with these Generative AI systems in natural language to express their intent. Generative AI systems interpret user intent and generate responses dynamically. Unlike traditional software, these responses vary for the same user inputs. This flexibility enables powerful capabilities but also introduces challenges such as hallucinations, inconsistent outputs, and potential exposure to biased, unsafe, or sensitive content. Hence, there is a need to design Generative AI systems such that human users not only learn to effectively specify what they want but also understand how to interpret and validate system responses.

Generative AI technologies have transformed how humans interact with technology. Instead of programming predefined instructions, users now express intent in natural language, and the AI technology generates outcomes. However, the outputs generated by Generative AI systems are often unpredictable and can vary significantly even with the same input. It can also hallucinate and produce plausible but untrue results. This paradigm shift therefore demands new design principles that provide guidance on how to design user experience for Generative AI systems and use them effectively and safely.

Generative AI systems deal with *generative variability, uncertainty, and evolving models.* Hallucination, toxic language, copyright infringement, and privacy breaches are some of the challenges of interacting with Generative AI. User experiences must, therefore, be intentionally designed to guide users, communicate uncertainty, and provide safeguards that support safe and effective interaction. Organizations should apply a new set of UX design principles tailored for Generative AI—focused on three core pillars: trust, transparency, and adaptability.

Design Responsibly: Technology with Purpose

Generative AI applications should be designed responsibly to solve real user problems and minimize user harm. Responsible design begins with asking: *What problem are we solving, for whom, and with what impact?*

Designers must adopt a **socio-technical mindset** and question how Generative AI will improve the user's experience, provide new capabilities, or address pain points for the users. Applications should be designed with a human-centered approach, such as design thinking and participatory methods, to understand users' workflows and real pain points. The proposed uses of Generative AI must align with users' actual needs, not for the sake of adopting technology. It is also important to consider the needs of all stakeholders—users, developers, and organizations—and strike a balance to avoid any harmful outcomes. Responsible design helps to reduce the potential for misuse and unintended consequences, addressing the ethical concerns around Generative AI.

Key Takeaway: *Responsible design isn't a constraint—it's a competitive differentiator. It builds trust and longevity into your AI systems.*

Design for Mental Models: Make AI Understandable

Users approach Generative AI systems with a preconceived notion based on their past software experiences. However, Generative AI behaves differently and can produce different responses for the same user input that breaks user expectations.

A mental model is the understanding of people of how something works and how its actions affect them. It is a simplified representation of the world that people use to process and apply new information. This principle promotes user-centric design and helps users interact effectively with the system. Users must know and understand that the output of Generative AI applications can vary for the same user input.

The application must provide in-context examples and explanations to help the user understand how to use the AI system effectively. For example, a health app should explain why certain suggestions were made.

Applications like DALL-E provide a curated set of example images along with the prompts used to generate them. This helps users understand the system's behavior and output.

Key Takeaway: *Bridging the gap between AI reasoning and human intuition builds credibility and accelerates adoption.*

Design for Appropriate Trust and Reliance

Trust is a critical factor for all Generative AI systems. Outputs from Generative AI systems should be accompanied by enough supporting information and explanations to help instill human trust. AI systems must transparently **communicate their capabilities and limitations**. This transparency helps users to better judge when to accept or question the AI recommendations. It should also teach humans when to be skeptical and scrutinize the outputs for bias, inaccuracies, quality issues, under-representation, or other issues, and reject and request modification, if required. Instill human trust in the output by providing **supporting evidence and source citations** for the generated output. For example, Google Gemini and Perplexity provides a list of sources they used to produce answers to questions. Determine and specify the role that the Generative AI system plays in the overall workflow. For example, GitHub's Copilot says that it is an AI pair programmer, which elicits its role as a partner. Embed **confidence scores and** quality indicators for users to rely on the output. Also, provide mechanisms for users to provide feedback on the AI response.

Key Takeaway: *Trust is built through transparency—show your work and users will trust the outcome.*

Design for Generative Variability: Embrace the Unpredictable

Generative AI systems can produce diverse and distinct outputs for the same input. These outputs may vary in character and quality, which is both a challenge and a strength. Some or all of these outputs may be satisfactory for the user. Because of this variability, it is difficult to replicate the results when working with Generative AI systems. The system must be designed to help the user identify how the outputs differ and to guide them in making informed choices. This variability must be **managed, not eliminated** with the following approaches:

- Allow users to organize, label, and filter AI outputs.

- Allow customization of creativity levels, tone, or randomness.

- Encourage iterative exploration—making variability a feature, not a flaw.

Key Takeaway: *Generative variability is a strength and not a flaw. Treat it as a feature, but add guardrails so users can reach a stable final answer.*

Design for Co-Creation: Empower the Human-AI Partnership

Generative AI thrives when humans and AI collaborate. Enable users to collaborate and co-create with Generative AI systems to **amplify human creativity and expertise**. Enable users to refine and enhance AI outputs collaboratively for a better experience. The system can guide on how to prompt effectively and improve their instructions to produce outputs that fit their needs. Users should also be able to define constraints and preferences, such as the number of outputs, and also co-edit with AI to improve the generated output.

Key Takeaway: *Co-creation shifts Generative AI from being a tool to being a collaborator—one that accelerates human potential.*

Design for Imperfection: Acknowledge and Address Errors

Generative AI systems may produce results that are incorrect or inaccurate, not aligning with user expectations, such as visual misrepresentations in images, bugs, or errors in source code. The application must warn the user about such imperfections, identify detectable uncertainties or flaws, and provide built-in recovery paths. For example, Google's Gemini includes a disclaimer that alerts users about uncertainties or imperfections in its outputs. Generative AI applications must also offer ways for users to correct or improve the AI's suggestions. Users should be able to provide feedback, regenerate content, or adjust results to better align with their goals.

Design for Human Control: Keep Humans in the Loop

Human beings must always be in control of the AI application and remain the **final decision-maker**. Generative AI applications must be built to augment human capabilities and not to replace or overpower them. The speed and creativity of AI systems must be balanced with human judgment, ethics, and expertise.

Prioritizing human control is essential for building robust, trustworthy, and high-performing Generative GenAI systems. Human control can identify and correct biased results before they cause any reputational, legal, or financial damage. Executives must define which Generative AI applications are allowed to perform, and which are not, without human approval. Such ethical guardrails help to prevent risky outputs. All outputs produced by AI applications must be critically reviewed by humans before being published and must be marked as "AI-generated." In all cases, users must have direct control over AI outputs. Feedback mechanisms should be built into the application for users to report errors and refine outputs for quality control and improvements.

Key Takeaway: *AI should augment human judgment, not replace it. Human control is the foundation of trustworthy AI.*

Design Generative AI Systems Iteratively: Learn, Improve, Evolve

Generative AI systems must evolve based on user feedback and usage patterns. An iterative approach to building Generative AI systems ensures that user feedback is continuously incorporated into the design process, leading to systems that evolve based on real-world interactions and user needs. Start building the application with a basic prototype focusing on the key user needs, aligning with the key design principles. Deploy and test it with a small group of users to gather feedback on usability, effectiveness, and personalization. Subsequently, refine the system based on the feedback received, addressing areas that didn't meet user expectations. Repeat and scale to add new features.

Key Takeaway: *AI design is never done—it's a living system that learns from every interaction.*

Four Layers of Generative AI Architecture

With Generative AI, humans specify the "what" and AI system decides the "how." A robust Generative AI architecture must ensure **scalability** to handle massive workloads, **trustworthiness** for responsible AI adoption, and **sustainability** for the future. The output of Generative AI is probabilistic and not deterministic—making explainability, transparency, and a feedback loop essential. The data powering the systems must be secured and continuously refreshed. The infrastructure must be scalable and cost-effective to support future growth. The governance must focus beyond code to review the model, data, and prompts. This brings in a fundamental shift in how Generative AI systems are designed, built, integrated, and governed. It requires a layered modular architecture to support **agility, reliability, and responsible innovation.**

The following four layers collectively form the foundations of Generative AI architecture:

- **Foundation Layer:** Consisting of infrastructure, cloud, networking, platform, and data

- **Intelligence Layer:** Covering the model life cycle management and operations

- **Integration Layer:** Consisting of applications, APIs, workflows, and user experience

- **Governance Layer:** Ensuring security, transparency, ethics, trust, and observability

Together, these layers build a resilient, modular, and scalable ecosystem for developing and deploying Generative AI solutions. Figure 6-1 shows the architecture stack details for a Generative AI application.

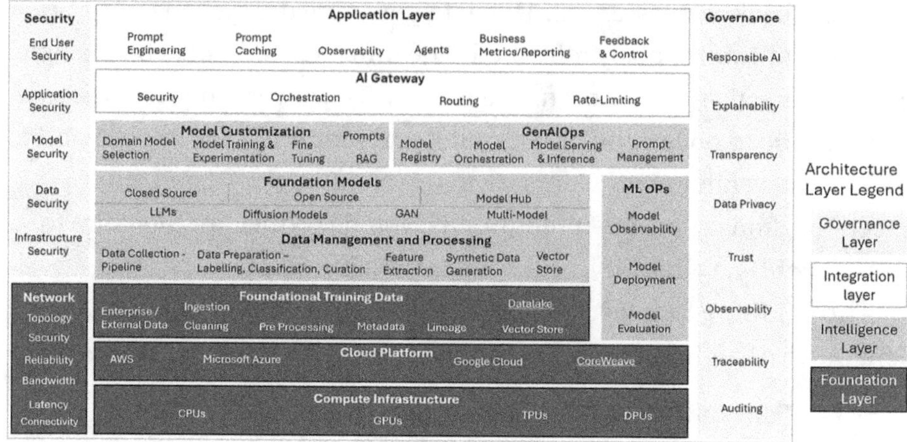

Figure 6-1. *Architecture Stack for Generative AI*

The subcomponents of each of the main pillars/layers of the Generative AI architecture are as follows:

Foundation Layer

Compute and data are core strategic assets that determine an organization's AI readiness and ability to scale. This layer gives every AI workload the computational power and data capabilities needed to succeed.

- **Compute Infrastructure** is the foundation of the entire AI ecosystem. It provides the raw computational hardware needed to run high-performance compute operations for model training and inference.

- **Cloud Platform** provides managed, auto-scalable container clusters and serverless compute for running AI workloads. A hybrid and multi-cloud setup combine on-premises control with cloud elasticity.

- **High-bandwidth network** layer provides a secure, low-latency network for seamless communication between compute nodes and for high-speed data transfer needed for distributed model training and inferences.

- **Foundational Training Data** provides the data, storage, and processing pipelines needed for training the foundation models.

Intelligence Layer

This layer focuses on the intelligence engine that powers Generative AI applications. It covers the entire life cycle of Generative AI: selecting the foundation models, fine-tuning them with additional domain-specific data, and finally using MLOps and GenAI Ops to automate and govern the training, deployment, serving (inference) in production.

- **Foundation Models** are the open-source and closed-source models that power the Generative AI application.

- **Data Management and Processing** provide pipelines and frameworks to ensure high data quality for training and experimenting with Generative AI models. This layer ensures that data is accessible and compliant at every stage of the Generative AI life cycle. It provided data ingestion, pre-processing, cleaning, and storage capabilities. Platforms such as Databricks, Snowflake, and BigQuery are generally used to unify and govern large-scale AI data ecosystems.

- **Model Customization** adapts pretrained foundation models for industry-specific needs to meet business, domain and regulatory requirements. Techniques such as fine-tuning, parameter-efficient tuning, retrieval-augmented generation, and prompt engineering are used to adapt and customize models for business contexts and specific industry use cases.

- **MLOps** provides a framework and tools for reliably training, deploying, and monitoring AI models at scale while maintaining quality, security, and governance.

- **GenAIOps** extends MLOps to Generative AI by addressing unique needs like prompt management, model orchestration, retrieval-augmented generation, and continuous fine-tuning.

Integration Layer

This layer covers the applications and the workflows built using the Generative AI models to deliver business value and transform consumer experiences

- **AI Gateway** centralizes and secures access to models, manages prompts, and enforces governance, security, and usage policies, ensuring scalability, observability, and consistent performance.

- **The Application Layer** provides the user-facing experience by integrating models with business logic, workflows, and interfaces using APIs and microservices. Workflow orchestration connects AI outputs to enterprise systems like CRM, ERP, HRMS and combines predictive and Generative AI to ensure that AI outputs are contextual, actionable, and aligned with the real needs of users and enterprises. It also captures user feedback for iterative improvement.

Governance Layer

This layer addresses the new risk associated with Generative AI. It helps to protect the data, models, and applications from data leakage and theft. It provides the guidelines and principles to build safe, ethical, and trustworthy applications that are unbiased.

- **Security layer** safeguards Generative AI applications by protecting data, models with a zero-trust data protection, and model isolation approach. It also prevents misuse, breaches, and adversarial attacks, ensuring compliance, trust, and resilience by ensuring that security and privacy are embedded throughout the stack.

- **Governance** covers the principles needed for building responsible, ethical, and trustworthy Generative AI solutions that can promote the safe adoption of AI at an enterprise scale. Human-in-the-loop design to validate AI outputs ensures safety.

Generative AI Infrastructure: The Engine Behind Intelligence

Generative AI infrastructure consists of the compute, storage, network and platform services required to build, train, deploy, and manage Generative AI models and workloads. A robust foundational infrastructure is necessary to train complex models, process multimodal data, and deliver low-latency inference at scale. Powerful and specialized computing resources like GPUs and TPUs, high-speed networking, and special AI services and tools are generally needed. A well-architected Generative AI infrastructure must balance the following:

- **Performance** to handle massive compute and data workloads

- **Scalability** to elastically handle the rapidly growing workloads

- **Cost Efficiency** to ensure sustainability and predictable ROI

Components of Generative AI Infrastructure

The following are the main components of Generative AI infrastructure:

Hardware for Parallel Processing

The foundation of AI infrastructure is the hardware that powers AI computation and accelerates AI processing. The infrastructure must be scalable and performant to handle the parallel processing of large training datasets. It should also handle the high resource needs for inference and reasoning.

New architectures like **Cerebras WSE, AMD Instinct,** and **Groq Chips** are reshaping compute economics by offering competitive throughput and lower power consumption than traditional GPUs.

High-Speed Networking

As model parameters and datasets grow, high-speed networking becomes as critical as compute. A high-speed, low-latency, and high-bandwidth networking solution is necessary for smooth communication between compute nodes and fast data access and storage needed for training foundation models and running AI workloads. Some of the architectural requirements for high-speed networking for Generative AI workloads are as follows:

- Ultra-Low Latency of < 10 microseconds interconnect (e.g., InfiniBand, NVLink)

- High-bandwidth of 400 to 1600 Gbps to handle multi-terabytes of datasets (e.g., 1600 Gbps of Elastic Fabric Adapter [EFA] from Amazon and Mixture of Experts [MoE])

Specialized AI Accelerator

Generative AI workloads are compute intensive. They require substantial power and memory to train models, handle large data volumes, and make ongoing inferences. GPUs are well-suited for parallel processing multiple operations simultaneously. There is also a growing adoption of specialized accelerators (ASICs, TPUs, FPGAs, etc.) and innovations in interconnect technologies (e.g., NVLink, CXL, high-throughput networking) to handle multi-GPU model training and distributed workloads.

Data Infrastructure

Data processing for Generative AI involves data collection, cleaning, processing, and transforming the data before feeding it into the training dataset. High-speed data processing requires a robust data infrastructure, including data lakes, warehouses, and vector databases, to ensure quality and accessibility. The core components of data infrastructure are

- **Unified Data Lakehouse** that combines the flexibility of data lakes and the structure of a warehouse for data storage.

- **Vector Databases** for retrieval-augmented generation for contextual data in responses.

- **ETL and DataOps Pipelines** to automate data preparation, cleaning, processing, labeling, and versioning.

Platforms such as Databricks, Snowflake, and Datadog support these needs. Databricks offers a unified platform for data engineering, collaborative data science, and business analytics. Snowflake offers a data warehouse solution specifically designed for the cloud, enabling secure and efficient analysis of large datasets. Datadog monitors and analyzes data across cloud platforms.

Cloud Platforms and Services

Enterprises often prefer on-premises infrastructure for data sovereignty. However, existing on-premises infrastructure often has limitations that drive the adoption of cloud platforms and services. Cloud providers like Google Cloud, Microsoft Azure, and AWS offer flexibility and scalability with scalable computing power, storage, networking, and pre-built AI models and tools for integrating Generative AI into applications. On-premises infrastructure is suitable when data sovereignty and security are primary concerns, but it requires significant CAPEX and offers limited flexibility. In most cases, a hybrid /multi-cloud setup is the preferred strategy as it balances control and agility. Hybrid infrastructure gives the enterprise the freedom to scale AI where it makes the most sense—close to the data or close to the user.

Key Challenges

Infrastructure Availability and GPU Scarcity

As demand for Generative AI workloads increases, the need for specialized GPUs grows. All major hyperscalers are seeking to significantly expand their GPU supply. With NVIDIA currently dominating the GPU supplier market, the wait time for its leading-edge chips is long. Hence, everyone is looking for alternatives to GPU chips or ways to access NVIDIA's GPU capabilities via new delivery models. Some alternatives that have emerged from this race include new GPU offerings from AMD and Cerebras, as well as AI ASIC chips like TPUs from Google and Inferentia chips from Amazon. Even new chip architectures, such as those from Etched and Groq, have emerged, especially for LLM inference. Various inference optimization libraries, such as CentML and OctoAI, also help reduce infrastructure requirements.

Data Quality, Security, and Compliance

The infrastructure must protect intellectual property and customer data. Data must be curated, de-duplicated, and anonymized before it is used for training or fine-tuning. Ensuring the security of large datasets, which are often used for training and fine-tuning AI models, is a key challenge. Handling fragmented, diverse data across platforms and business units to achieve a unified view and a single source of truth requires a scalable, secure data infrastructure. Organizations also need to handle data sovereignty requirements while balancing network latency when accessing data.

Some of the best practices for data handling are as follows:

- Enforce **zero-trust data** policies and lineage tracking

- Implement **PII redaction** and **data tokenization**

- Ensure compliance with GDPR, HIPAA, and upcoming AI-specific regulations (like the **EU AI Act**)

Cost Management

Training, fine-tuning, and inference of AI workloads can be a minefield of unexpected expenditure. The computational power and storage required can lead to high operational costs. Cloud providers can offer cost-effective options, such as spot instances, reserved instances, and pay-as-you-go solutions, to help avoid over-provisioning. Using AI-optimized hardware like GPUs and TPUs, and a software stack like TensorFlow and PyTorch can help to minimize resource wastage, maximize hardware utilization, and optimize IT spend. **Parameter-efficient fine-tuning (LoRA, QLoRA)** can cut training costs. Use AI workload schedulers to dynamically allocate resources based on utilization. All such optimization techniques can lead to substantial cost savings.

Power, Space, and Cooling

Large-scale Generative AI solutions require significant power and a robust cooling infrastructure. Building and managing such an infrastructure is complex and resource intensive. Hence, organizations are increasingly using cloud service providers and independent software vendors for their Generative AI needs.

Infrastructure Adoption Trends

Hybrid Cloud AI Infrastructure

Most enterprises are moving toward a hybrid model. A hybrid approach that combines on-premises and cloud infrastructure (multi-cloud or single public cloud) provides maximum flexibility and cost efficiency. This approach enables organizations to select the optimal environment for each workload, balancing security, cost, and performance requirements. It also meets the regulatory requirements regarding data residency, privacy, and compliance.

Many financial institutions train their models in an on-premises infrastructure for data privacy and deploy inference on a public cloud for scalability.

Leverage Proprietary Models

Using proprietary models like Gemini, Claude, and GPT helps prioritize speed, ease of use, cost-effectiveness, and scalability. Building custom AI models is expensive and time-consuming and should be considered only for special use cases where proprietary models cannot deliver the required output. Managing custom models needs deep expertise in AI, data engineering, infrastructure, model training, fine-tuning, monitoring and governance. Most organizations prefer to use proven foundation models as they allow them to focus on business problems and deliver value instead of investing heavily in building and maintaining AI models from scratch.

Containerization and Orchestration of AI Workloads

Generative AI services not only have high computational and memory requirements but also require dynamic scaling to meet varying demands. Managing software dependencies and version control to ensure consistency of Generative AI models across different environments adds to complexity. Distributing computational load across all available resources is vital to ensuring the stability and performance of AI services.

Containerization of AI workloads simplifies the deployment and improves portability. It ensures consistency and reduces the likelihood of conflicts between dependencies of the AI models used. MLflow, Kubeflow, and Vertex AI are popular frameworks for container-based ML life cycle management.

Container orchestration with Kubernetes enables easier updates and auto-scaling. It simplifies the roll-out of updates, thus reducing operational complexity. It also helps to optimize resource utilization and operational costs of Generative AI systems.

Containerization makes AI portable—Kubernetes makes it scalable.

Platform Engineering for Generative AI: From Experiments to Enterprise Systems

Why Platform Engineering Matters

Developing and scaling a Generative AI application is inherently complex, often leading to lost productivity, irreproducible experiments, and deployment errors. Some of the factors that contribute to this complexity are

- **Model Churn**

  - The foundation models are continuously evolving at a fast pace

- **Data Readiness**

  - Gaps in data quality and data democratization

  - Inefficient data pipelines impacting model fine-tuning

- **Life Cycle Automation**

  - Absence of robust automation for model life cycle management

  - Lack of coordination between the data scientist, development, and operations teams for model deployment

  - Versioning chaos—difficulty tracking datasets, prompts, and model checkpoints

- **Governance and Monitoring**

  - Inconsistent governance and poor visibility across teams

  - Lack of continuous monitoring of models in production to guard against drift

Many organizations fail to realize the ROI of their investments in Generative AI projects because most of those projects never move to production. A well-designed **AI platform** solves these challenges by abstracting infrastructure complexity, standardizing processes, and automating repetitive tasks across the AI life cycle.

However, the growing number of tools to address different challenges has introduced significant complexity in choosing, integrating, and operationalizing the right platform and framework from the vast array of options available. *Without a cohesive foundation, organizations risk building siloed solutions that are hard to scale, govern, or sustain.* This is where **Platform Engineering for Generative AI** becomes essential. By establishing a standardized MLOps and GenAIOps platform, organizations can automate and accelerate the end-to-end AI life cycle, improve collaboration across teams, optimize cost and performance, and ensure governance and reliability in production.

Key Takeaway: *Platform engineering transforms Generative AI from experiments with isolated models into enterprise-grade capabilities that automate, orchestrate and govern intelligence at scale to deliver measurable business outcomes.*

6 Layer Generative AI Platform Stack

- **Collaborative Development Environment**

  Standardized workspace for data scientists, ML Engineers, and developers to build, experiment, and collaborate. Tools like Jupyter, VS Code, and Azure ML Studio offer pre-configured environments with dependency management and shared tooling that simplifies and streamlines the machine learning ecosystem.

- **Integrated Data Engineering Pipelines**

  Automate pipelines for data ingestion, validation, cleansing, transformation, and cataloging that ensure a trusted production-ready dataset for accurate AI outcomes.

- **Model Development and Training Environment**

  Scalable environments to build, train, fine-tune, and experiment with models with training data. Frameworks such as Tensorflow and PyTorch are used for building, training, and deploying machine learning models.

- **CI/CD and Release Automation Framework**

  A robust and scalable delivery framework that supports automated testing, packaging, deployment and rollback of AI Models for improved resilience.

- **Observability and Guardrail Intelligence Layer**

  Continuous monitoring of model and system behavior to track latency metrics (response time, throughput), cost metrics (token usage, inference cost, cost per workflow), quality metrics (accuracy, hallucination, human override rate) data drift metrics and bias detection. Tools like Prometheus, Grafana, or Datadog can be used for performance visualization and alerting.

- **Governance and Access Control Framework**

  Enterprise policies, role-based access controls, audit trails, and responsible AI guardrails to enforce security, compliance, and accountable usage.

The Evolution from MLOps and GenAIOps

A robust MLOps platform is essential for deploying Generative AI solutions to production more quickly and at scale, with the objective of reducing the time it takes to take a Generative AI use case from concept to production. This helps businesses stay competitive and adapt to evolving market demands.

Traditional **MLOps** builds on top of DevOps and focuses on the following aspects of the model life cycle:

- **Model Packaging** bundles the model with all the dependencies. Models can be containerized as Docker containers, including all dependencies, to ensure consistency across different environments. Model management platforms like MLFLow can manage the ML life cycle and help in packaging and deploying models in various formats.

- **Model Serving** sets up the infrastructure and makes the trained model ready to receive and respond to requests with the predicted results. A CI/CD platform like Jenkins, GitLab CI/CD, or Azure DevOps can be used for training, testing, and deploying the model. Kubernetes can automate the deployment, scaling, and management of containerized models.

- **Monitoring and Logging** track the health and performance of deployed models. Tools like Prometheus and Grafana can be used to monitor metrics and visualize model performance. Logging all events and data can help detect anomalies in the model behavior over time, ensuring reliability and transparency of Generative AI services.

- **Scaling and updating** manage the resources to handle the varying load and update the model with the latest changes. Cloud-based autoscaling platforms and load balancers can help ensure model scalability under high load.

Model life cycle management, which tracks the versions of models, datasets, and outputs throughout the life cycle, enables traceability and rollback if bias or drift occurs. A model registry facilitates version control and storage of models. It connects the experimentation phase (where a data scientist trains the model) with the operational phase (where the model is deployed and monitored).

GenAIOps extends MLOps with new operational disciplines that address the unique challenges of Generative AI. Generative AI builds upon the pre-trained foundation models that respond to user prompts.

The responses vary based on the prompts. Hence, **prompt engineering** is a crucial aspect for achieving accurate, optimal results. Managing these prompts effectively forms the discipline known as **PromptOps**.

Retrieval Augmented Generation (RAG) further enhances the responses with more contextual and enterprise-specific information to address specific business needs. Fine-tuners adapt foundation models to specific tasks using enterprise-specific data. This involves managing the fine-tuning process, retraining models (or parts of them), and ensuring optimal performance—all of which constitute **RAGOps**. Similarly, **AgentOps** focuses on managing the operations of the autonomous Generative AI agents.

Together, PromptOps, RAGOps, and AgentOps form the broader practice of **GenAIOps**. Figure 6-2 shows how all these are interconnected.

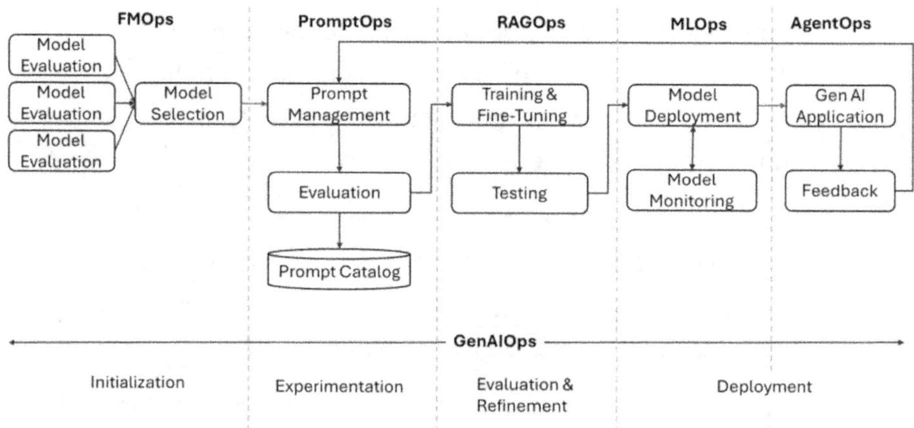

***Figure 6-2.** GenAIOps Flow from Initialization to Deployment*

Without a structured GenAIOps framework, organizations often struggle with

- **Model drift and hallucination** due to a lack of feedback loops.

- **Uncontrolled costs** from inefficient compute scaling.

- **Security and privacy risks** in data handling can lead to prompt exposure.

- **Ethical and compliance challenges** due to unmonitored model outputs.

GenAIOps extends MLOps with more specialized operational frameworks that enable organizations to automate and manage the full life cycle of Generative AI models—from data preparation and training to deployment, monitoring, and continuous improvement.

GenAIOps fosters collaboration among teams of data scientists, prompt engineers/testers, and AI engineers. ML engineers fine-tune generative models and address ethical concerns by implementing frameworks for identifying and mitigating potential biases inherent in generative models. Ultimately, GenAIOps also captures user feedback and integrates it into retraining or fine-tuning cycles to drive continuous improvement and enhance contextual relevance.

Security Architecture for Generative AI

The most significant risks associated with Generative AI are the exposure of sensitive data through prompts, training data, model outputs, or integrations. Careless human practices, when interacting with Generative AI tools, can accidentally leak sensitive data through input prompts and AI outputs. A robust security architecture is important to redact PII, enforce strict access control, and filter prompts/outputs with audit logging. An adaptive and layered security approach must be implemented, with AI-specific controls for infrastructure, data, models, and applications.

Core Layers of Generative AI Security Architecture

Infrastructure Security

Generative AI requires specialized hardware for running large language models with low latency. This infrastructure must be secured from unauthorized access by any malicious actors. The following are some of the approaches that can be used to secure the infrastructure:

- **Network Segmentation** to strictly isolate and secure the infrastructure hosting the model training and serving clusters, RAG vector databases, and the training datastores from the public-facing application.

- **Model Isolation** using containers to deploy and isolate the serving model from the rest of the application network can limit the impact if the models are exploited or compromised.

- **Authentication and Authorization** using Multi-Factor Authentication (MFA), API gateways, and Role-Based Access Control (RBAC) can be used to secure and restrict the API access to the models.

- **Edge protection** using a Web Application Firewall and DDoS can be used for input sanitization to filter malicious prompts or code injections.

Data Security

Data used for training and fine-tuning AI models must be secured to prevent incorrect data from being used, as this can affect model performance and outputs. The following are some of the approaches to ensure data security:

- **Encrypt data** at rest and in transit using strong algorithms during training and inference stages.

- **Anonymize or tokenize sensitive data** before it's used for training or fine-tuning to mitigate privacy risks.

- **Apply zero-trust policy and the principle of least-privilege access** to secure all data stored in the vector database. Secure training datasets and inference logs from unauthorized access.

- **Implement data governance with continuous lineage** tracking to meet regulatory obligations such as GDPR and the EU AI Act.

Model Security

AI foundation models and fine-tuned models must be protected against poisoning and tampering. The following are some of the approaches that can be adopted for the same:

- **Use version control, digital signatures, and cryptographic checksums** to protect the integrity of the models and ensure they have not been tampered with.

- Implement input **validation and sanitization** to detect and block malicious prompt inputs.

- Implement **rate-limiting and API-level validation** to ensure model inference endpoints are not compromised by malicious actors.

- Implement strict policies to validate data and detect anomalies before the data enters the model training pipeline.

- Conduct **"AI Red Teaming"** exercise with security experts trying to breach the model's security in an attempt to discover and fix vulnerabilities in the model before hackers do.

Application Security

Generative AI applications must be secured with a multi-layered, zero-trust security approach to protect them from misuse and compliance issues. The following are some of the recommended approaches to secure the application interface:

- Protect chat UI and backend APIs using user authentication and OAuth/JWT Tokens, respectively.

- Use Role-Based Access Control (RBAC), multi-factor authentication (MFA), and secure API gateways to authenticate and authorize access.

- Implement continuous, real-time logging and monitoring of user prompts, model outputs, and resource utilization for complete observability.

Other Architectural Considerations for Scaling Generative AI

Recommend Options and Predict Impact to Enhance Human Decisions

AI is evolving from making predictions to assisting humans in making informed decisions. Combining generative and predictive AI can empower human decision-making. It helps to create, refine, and present options to human decision-makers. Different choices presented by AI empower humans to exercise meaningful judgment and think strategically, enabling them to make the right decisions that matter most.

Gradually Build Trust in AI with Human-In-The-Loop

For Generative AI solutions to scale and be widely adopted, they must be trustworthy. Building trust is a gradual process. This involves first building trust in AI with low-risk decisions before it can be adopted for high-risk processes. It also needs structured organizational training and a culture that evolves with change.

Implement Guardrails for Resiliency

Guardrails for Generative AI Apps are rule-based filters that guide or constrain the output of Generative AI. The intent is to keep the AI output safe, ethical, and aligned with organizational or societal standards. Guards and validators inspect input prompts and responses, enforcing application-specific rules to provide comprehensive protection.

Generative Model SLAs and Regional Availability

Many Generative AI models lack strong SLAs, creating reliability risks for production workloads. Regional model availability across geographies can impact latency and performance. Even model rate limits and quota restrictions can hinder scalability at high loads.

Conclusion

A resilient architecture and secure infrastructure form the foundations for a scalable Generative AI solution. An intuitive and human-centric UX design makes the application easy to use and drives adoption. A long-term value requires trust, governance, and adaptive security. Successful implementations combine collaborative development, continuous feedback, and clearly defined guardrails to ensure responsible and ethical use. With these foundations in place, enterprises can unlock the potential of Generative AI to have a safe, creative, productive, and positive impact on industry and society.

CHAPTER 7

Building the Data Architecture for Generative AI

Introduction

Data is driving innovation, decision-making, and competitive advantage across industries. High-quality data is the foundation for all Generative AI systems. Data is needed not only for training foundational models, but also for fine-tuning pre-trained models to incorporate domain-specific information. RAG implementations also need complete and accurate data to provide reliable responses. Data needs to be collected from reliable sources and pre-processed to validate its quality and accuracy, and then fed into the Generative AI models and vector databases. All of this requires strategic planning and a robust, scalable data architecture that can efficiently and securely process and store massive volumes of data in varied formats. The architecture must detect and mitigate bias. It must provide visibility into data lineage and follow sustainable, cost-effective data processing procedures.

© Brajesh De 2026
B. De, *Generative AI for Business Innovation*, https://doi.org/10.1007/979-8-8688-2682-5_7

This chapter will focus on the approach and best practices for building a robust and modern data architecture that supports powerful, reliable, and responsible Generative AI systems. It explores five key elements of Generative AI data architecture: data lifecycle management, data strategy, governance, security. and readiness assessment.

Reference Data Architecture for Generative AI

Generative AI requires a scalable data architecture capable of handling large volumes of diverse data types. The architecture should be modular, enabling it to connect to and collect data from different sources and feed that data into different models. The architecture should provide real-time data processing and scalable storage to meet the target latency and throughput requirements of any Generative AI application.

Some of the key architecture capabilities needed for data processing for Generative AI are

- Data Ingestion from varied sources in various formats

- Efficient and scalable storage

- Real-time data pipelines to support continuous learning

- Streaming architecture for high-volume, low-latency data processing

- Semantic search using vector databases for RAG implementations

- Specialized indexing and query optimization

- Data privacy and security for regulatory compliance

Figure 7-1 shows the key components of data architecture for Generative AI.

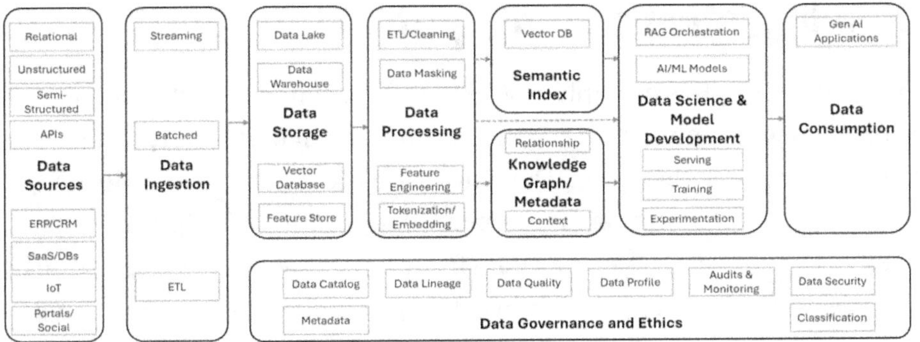

Figure 7-1. *Key Components of Data Architecture for Generative AI*

The data architecture for Generative AI must address the following key considerations:

- Handle massive and diverse datasets scalably

- Store, retrieve, and process data efficiently

- Ensure security and privacy of sensitive data

- Manage data quality and mitigate bias

- Track version and lineage of the data

- Must follow cost-effective, sustainable data management practices

The Generative AI Data LifeCycle

- Data Ingestion and Integration

- Data Storage and Access

- Data Processing and Transformation

- Data Retrieval and Augmentation

- Data Monitoring

Data Ingestion and Integration

Every enterprise has multiple systems, each with its own data source. Siloed data has limited value. Hence, data from all these sources need to flow into a unified data layer. Data ingestion is the process of collecting, preparing, and loading data into a central repository. Multiple systems need to be integrated to extract data from sources such as databases, files, streams, and APIs, and load it into a data warehouse, data lake, or other data storage infrastructure. For Generative AI, the data ingestion and collection strategy must also address the various data modalities.

Even the complexity and scale of data integration have increased as the number of IoT devices, streaming technologies, and connected systems has grown. Depending on the business scenario and the Generative AI use case, data may have to be refreshed and synchronized either in real-time, near real-time, micro-batch, or batch. For example, customer support assistants or fraud detection systems may require near real-time data to respond accurately and promptly. On the contrary, strategic reporting systems can rely on data available in batches or in periodic intervals. The data freshness requirements will determine the appropriate data integration architecture.

The data ingestion and integration pipeline must be designed to address the following:

- Integrate with existing systems and diverse cloud or on-prem data sources

- Handle large volumes of data of diverse data types and modalities

- Extract relevant information from unstructured data, such as text, images, audio and videos

- Support both real-time and batch ingestion of data

- Ensure privacy and security of data

- Transform data from an unstructured to a
  structured format

- Ensure quality and completeness of ingested data

- Ensure regulatory compliance for data integrity,
  confidentiality, and availability

- Track the source and lineage to ensure transparency
  and compliance

Data Storage and Access

Data storage systems for Generative AI need high performance, scalability, and volume. Data ingested from various sources needs to be stored. A data lake/Lakehouse is the preferred storage layer for storing raw data in its entirety. It is best to keep the raw data as long as possible, as it often provides the best source for model training. Cloud storage services such as Amazon S3 and Azure Data Lake Storage are recommended for raw data. Specialized vector storage like Pinecone, Weaviate, or Azure AI Search is needed to store indexed vector embeddings of data needed for RAG implementations.

Below are some of the architectural considerations for designing the right data storage architecture that provides scalability, secure access, and long-term storage:

- Meet the growing data storage needs while ensuring
  scalability and performance.

- Separate data into hot/warm/cold data buckets to
  optimize cost and access patterns for frequently and
  rarely used data.

- Design for multi-model data storage to handle text,
  image, audio, and video.

- Store data in both unstructured and structured formats.

- Provide fast semantic search through indexed high-dimensional vector embeddings.

- Availability of the storage for model training and inference.

- Ensure compatibility of older and newer data formats.

- Ensure the security and privacy of data both at rest and in transit.

- Provide lineage information and quality metrics for all datasets.

Data Processing and Transformation

Data processing and transformation ensure that the data is clean, relevant, and available in the right format for the model. The data must be of high quality, diverse, and complete. This ensures Generative AI outputs are accurate, relevant, and unbiased. Almost 70% of the model performance depends on this stage of data life cycle.

Data processing for Generative AI involves the following steps:

- Cleaning the data to remove duplicates and inconsistencies, and additionally handling missing values, low-quality content, and toxic data. Data cleansing is necessary to mitigate biases.

- Masking sensitive and PII data for privacy and compliance.

- Standardize and deduplicate data for consistency. Fix errors to improve data accuracy.

- Tokenization to transform raw data (text, image, etc.) into numerical tokens.

- Embedding generation helps AI models mathematically understand the meaning and relationships in complex real-world data, such as words, sentences, images, and audio.

- Add metadata, embeddings, and contextual tags to make the data ready for models.

- Feature Engineering to create high-quality features for fine-tuning and specialized models.

- Data Splitting to partition data into large training sets, validation sets, and test sets.

Data Retrieval and Augmentation

The LLMs are mostly trained on generic and publicly available data. But they do not have any knowledge about enterprise data. This is when these models need to be augmented with additional enterprise- or domain-specific information stored in vector databases. Embedding models convert enterprise documents, databases into vector embeddings that facilitate rapid similarity search. Retrieval-Augmented Generation (RAG) is a core data architecture pattern that performs a semantic search in a vector database, retrieves the results, and then uses an LLM to respond to the user in the requested format. Unstructured content has to be chunked and indexed to improve search performance. Some of the key considerations for data retrieval and augmentation are as follows:

- Security and Privacy of sensitive data stored in vector databases

- Timeliness of data to respond with the most up-to-date information

- Cost

- Scale

- Quality and Reliability

Data Monitoring

Monitoring the quality of the data used for training and fine-tuning Generative AI models and RAG implementations is important to ensure accurate and unbiased output. Proactively monitoring and identifying issues in data can avoid costly failures. Changes in input data distribution can impact a model's performance. Bias in training data can lead to unfair and biased outputs. Hence, it is important to constantly monitor the quality and drift in data.

Some of the key parameters to assess the quality of data are completeness, timeliness, consistency, uniqueness, validity, and accuracy. Statistical measures like Kolmogorov-Smirnov (K-S) test, Chi-squared test, Kullback-Leibler (KL) Divergence, and Population Stability Index (PSI) can be used to measure data drift. A change in model performance metrics like accuracy, precision, recall, F1-score, AUROC can indicate a change in underlying data distribution.

Five Pillars for Robust Generative AI Data Strategy

If Generative AI is a great meal, then data is the ingredient, and data strategy is the recipe that delivers top nutrition and flavor. Data strategy for Generative AI includes the policies, processes, architecture, and governance framework that collect, prepare, and deliver high-quality data to LLMs and Generative AI systems. You need a strong data strategy that delivers high-quality, transparent, and governable data. The Generative

AI data strategy must handle both structured and unstructured data. Unstructured data can be in diverse modalities—namely, text, image, video, audio. Additionally, the strategy must ensure quality and maintain semantic relationships between data from various sources to enable intelligent response generation.

The following are the five main building blocks for a solid data strategy for Generative AI as shown in Figure 7-2:

- Data Quality and Lineage

- Data Context and Relationship

- Data Ownership and Literacy

- Data Governance and Compliance

- Data Architecture

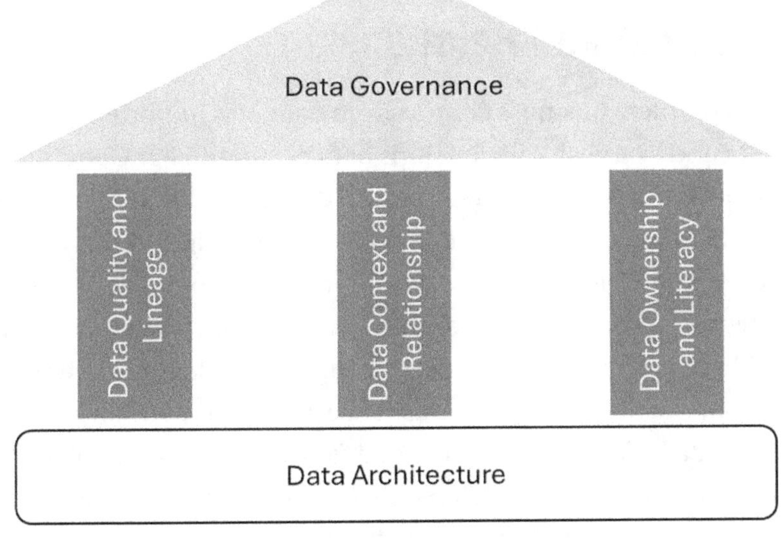

Figure 7-2. *Building Block of a Data Strategy for Generative AI*

Data Quality and Lineage

The Generative AI data strategy must ensure high-quality data for training and fine-tuning the Generative AI models. It must ensure that the data is complete, unique, up-to-date, consistent, and accurate. Poor quality of data can lead to hallucination, biased responses, and unreliable output. Lineage helps to understand the origin of the data and debug failures at the root level. High-performance Generative AI applications need the right data quality checks built at every life cycle stage of data flow and a clear lineage with source traceability. Semantic consistency, factual accuracy, and contextual appropriateness are some of the quality parameters to be focused for Generative AI.

Talend, Informatica DQ, and Great Expectations are some of the popular platforms for data quality management. Collibra Lineage, Manta, helps to provide end-to-end data lineage information.

Data Context and Relationship

To get the complete business context from data, it is important to connect siloed data and establish relationships between them. For Generative AI to respond with context and relevance, data from different business functions and domains must be connected. Knowledge graphs help to organize, integrate, and contextualize information from diverse sources. This represents a complex relationship between data entities and sources and establishes the semantic context to improve explainability and enable advanced interconnected reasoning between various data sources.

Siloed data lacks the context needed for a Generative AI system to respond. Without the context, Generative AI responses may be technically accurate but inappropriate for the business. Semantic relationships, data lineage, quality indicators, and business context must be part of metadata information for Generative AI. A structured network of nodes (entities) and edges (relationships) improves the ability of AI systems to understand, reason, and make informed decisions based on interconnected facts rather than isolated data points. Tools like OpenMetadata, Collibra, Alation, and Informatica provide a platform to catalog data, manage metadata, and map deep data dependencies and relationships.

Data Ownership and Literacy

Historically, data control has been either completely centralized or locked in silos across different business units. This approach hinders Generative AI from delivering full value without knowledge about the data context and usage. Hence, a federated data ownership model is needed to enable responsible data ownership and sharing across business units and teams. Define clear roles across the organization—domain data owners, product stewards, analysts—to promote accountability and faster alignment with business goals and Generative AI use cases. A centralized enablement team should provide governance, policies, standards, and oversight for the sharing and use of domain data.

Data literacy is needed to understand, curate, and assess data readiness for Generative AI use. Tools like Collibra, Ataccama One, and Informatica define, manage, and enforce who owns and is responsible for which data assets, increasing accountability.

Data Governance and Compliance

Data governance practices for Generative AI must promote responsible AI practices and ensure regulatory compliance. Training data and model outputs must be continuously monitored to detect bias or discriminatory patterns that could violate ethical standards or harm business reputations. The data governance framework should promote transparency by explaining and enabling understanding of how Generative AI systems make decisions. Content-filtering, output monitoring, and human oversight should be part of responsible AI practices for data governance.

Tools like Collibra, Alation, Microsoft Purview, and Informatica provide a comprehensive data governance platform with capabilities for data cataloging, lineage tracking, metadata discovery, and classification. Tools like Arthur AI and CalypsoAI help to secure and govern AI models to ensure fairness, detect hallucinations, validate prompts, and enforce policies for compliances.

Data Architecture

A unified and scalable data architecture is necessary for all successful Generative AI projects. The architecture must support diverse data types and formats—structured, semi-structured, and unstructured. It must support ingestion from various sources, which can be documents, slack messages, emails, code repositories, wikis, enterprise data sources, and more. The source data will need **pre-processing** to extract relevant information, break down large documents into smaller chunks while maintaining context and be cleaned to remove any sensitive PII or toxic data. After that, it needs to be transformed into a format that can be used by AI models through a process called **embedding** that converts a chunk of data into a vector representation that represents the semantic meaning of the text and contextual relationship. The vector embeddings need to be stored in **vector databases** and indexed into high-dimensional vectors for fast retrieval. The data architecture must support hybrid search that combines **Vector Search** (semantic meaning) with **Keyword Search** (exact matches) to ensure high accuracy.

Delta lake provides a lakehouse to support batch and stream data ingestion and raw data storage. Vector databases like Milvus, Pinecone, Weaviate, and ChromaDB provide high performance storage for storing vector embeddings used for LLM/RAG implementations. Lakefs/Data Version Control (DVC) provides the ability to version datasets to reproduce experiments reliably. Feature stores like Feast/Tecton help to manage and reuse features across training and inference. Kubeflow and Dagster orchestrate data pipelines, model training, and serving.

Data Governance, Compliance, and Ethics

Data governance, compliance, and ethics must define and enforce policies that ensure the ethical and responsible use of data for Generative AI applications. Clear protocols must be defined for data collection, storage, and usage. It must cover the following:

- Classify data and tag them by sensitivity, confidentiality, and usage rights.

- Enforce compliance with regulatory laws like GDPR, CCPA, EU AI Act, and DPDP and meet industry-specific requirements like HIPAA, FINRA, etc.

- Audit and monitor data for fairness and ensure that Generative AI models are trained and respond with unbiased data.

- Ensure ethical and responsible AI policies are followed to remove biases in data for fair and ethical usage of AI.

- Track the provenance, lineage, and traceability for GenAI data pipelines for explainability and transparency.

- Enforce policies to ensure data is used appropriately and prevent unauthorized use of sensitive data.

- Implement strong data quality checks with human oversight and continuous feedback.

Data Security and Privacy

The data used for training and fine-tuning Generative AI applications must be secured to protect all confidential information and also ensure the privacy of personal information. A zero-trust security architecture principle must be followed to ensure the safety and security of the data for Generative AI. The following best practices must be followed for data security and privacy:

- **Secure Data Access** using strong authentication mechanisms and role-based access control.

- **Ensure data confidentiality** using strong encryption algorithms to protect data at rest and in transit.

- **Anonymize personal data** by masking sensitive personal data used for model training.

- **Implement consent management** and usage tracking for personally identifiable information (PII).

- **Ensure Data Residency,** if required, by adopting a region-aware data architecture with cloud regions that comply with local laws. All storage, compute, vector DB, LLM endpoints should be configured in the same region.

Data Readiness Maturity Model

The data readiness maturity model measures the readiness of organization's data to support the training, fine-tuning, and operational use of Generative AI systems. It builds on top of traditional data maturity models to introduce additional capabilities required specifically for Generative AI. Beyond assessing how data is collected, stored, processed, governed, and used, it evaluates the **semantic readiness** (how well data is structured and contextualized for AI understanding), **preparation of unstructured content** (documents, conversations, images, and other non-tabular data), and the **embedding life cycle** (creation, storage, versioning, refresh, and monitoring of vector embeddings used in RAG systems) needed for Generative AI. This assessment is important as output of Generative AI is as strong and reliable as data feeding the models.

High quality data enables AI to make better decisions, improve customer experience, and detect anomalies in real time. Executives must recognize that limited data readiness leads to fragmented and inconsistent data that restricts their ability to deploy or scale AI. Figure 7-3 shows the five levels of data readiness maturity for Generative AI.

Low Maturity →→→ High Maturity

Maturity Level	Initial: Chaos	Developing: Organized	Defined: Foundational	Managed: Ready	Optimized: Native
Data State	Silo-ed, fragmented Inconsistent	Data inventoried, Basic structure and governance starting to develop	Foundational data architecture established and governed.	Data platform is optimized specifically for Generative AI	Fully intelligent, automated, self-optimizing data ecosystem
Characteristic	• Data stored in silos with inconsistent formats. • No unified view, no metadata, no data quality checks. • Poor lineage; unknown data provenance. • No defined processes for data access, labelling, or preparation.	• Data sources identified and inventoried. • Initial metadata management, catalogs, and glossaries in place. • Basic data quality rules applied for critical datasets. • Manual data cleaning, preprocessing pipelines created.	• Enterprise data models, semantic layers, and lineage implemented. • Automated data pipelines for ingestion, curation, and transformation. • Reliable master data; consistent identifiers across systems. • Basic governance committee and data ownership roles in place. • Structured + unstructured data unified for AI use.	• Vector databases, embedding pipelines, and feature stores operational. • Highly scalable, real-time data ingestion and semantic document processing. • High-quality labelled datasets for fine-tuning models. • Automated data quality monitoring and feedback loops. • Data governance tailored for AI: model lineage, prompt logs, data usage policies.	• AI agents auto-correct data quality. • Dynamic knowledge graphs with real-time semantics. • Continuous learning between apps, models, and data. • Governance automated through policies-as-code. • Synthetic data pipelines to boost training quality. • Seamless multi-modal data fusion across all formats.
Gen AI Capabilities Enabled	• Inconsistent, unreliable PoCs	• Early experimentation with vectorization and small-scale fine tuning	• Reliable RAG systems • Enterprise-wide pilot deployments with standardized embeddings	• Production-grade RAG, • Fine-tuning, multi-modal data flows, • Domain-specific model development	• AI systems consistently learn, adapt, optimize, and reason using near-real-time enterprise knowledge - enabling true intelligent automation

AI Data Readiness Maturity Levels

Figure 7-3. *Data Readiness Maturity Levels for Generative AI*

Data Architecture Checklist for CTOs and Architects

Below is a quick reference checklist for CTOs and Data Architects to ensure a robust and scalable data architecture that meets all the necessary requirements.

- **Data Strategy and Foundation**

  - Do you have defined data strategy aligned to Generative AI use cases?

  - Have the business domains, data owners, and consumers been identified?

  - Is the data acquisition plan for Generative AI defined?

  - Is the data readiness assessment completed?

- **Data Ingestion and Integration**

  - Are all source systems identified (structured, unstructured, multimodal)?

  - Are ingestion pipelines designed for both real-time and batch?

  - Is unstructured data handled (PDFs, images, videos, emails)?

  - Are ingestion policies enforcing PII redaction and sensitivity tagging?

  - Is lineage captured at ingestion?

- **Data Storage and Architecture**

  - Is there a central lake/lakehouse (S3, ADLS, GCS)?

  - Have hot/warm/cold storage tiers been defined?

- Has data storage format been standardized?

- Have data retention policies been defined

- Does the data storage architecture provide scalable access for RAG and fine-tuning?

- Is vector DB selected and schema defined?

- **Data Processing and Transformation**

  - Does a data catalog exist?

  - Have data quality checks been implemented?

  - Has toxic content filtering been implemented

  - Have bias detection algorithms and data drift checks been implemented?

  - Is the content chunking strategy finalized?

  - Is PII detection ready?

  - Is the embedding model selected?

  - Are metadata enrichment rules defined?

  - Is the training data labeled, cleaned, and version-controlled?

- **Data Governance and Compliance**

  - Are GDPR, DPDP (India), CCPA, HIPAA, and FINRA rules applied?

  - Is there an AI governance and Responsible AI policy defined?

  - Is data classification implemented (Public, Internal, Restricted, Sensitive)?

- **Data Security and Privacy**
  - Is data access control enforced with user authentication and RBAC policies?
  - Is encryption policy defined and applied to secure data at rest and in transit?
  - Is PII masked or tokenized before LLM ingestion?
  - Is Data sovereignty checked?
- **Data Residency and Sovereignty**
  - Is storage, pipelines, embedding generation, vector DBs, and logs region-bound?
  - Are LLM endpoints deployed in-region?
  - Is cross-region data movement blocked with IAM + network policies?
  - Are customer-managed keys located in-region?

Conclusion

Implementing the right data architecture for Generative AI is one of the most important steps for a successful Generative AI program. It is a strategic transformation that touches every layer of the enterprise ecosystem. Organizations must build strong foundations in data governance, quality, security, and scalability to enable sustainable Generative AI driven innovations.

Building a robust and scalable Generative AI data architecture starts from understanding the use case and identifying the data required to support it. The data architecture must seamlessly integrate with diverse data sources, process unstructured and multimodal datasets, and deliver high-quality, reliable information to AI models. Embedding pipelines, vector databases, retrieval-augmented generation (RAG) frameworks, and scalable storage all play critical roles in enabling accurate, contextual, and domain-aware responses.

Organizations must also invest in setting up a strong data governance framework by establishing ethical AI practices and ensuring regulatory compliance is met at every layer—from ingestion and processing to retrieval and inference. Ensuring data quality, securing sensitive data, ensuring privacy, managing data residency, and monitoring data drift and bias are some of the essential guardrails for setting up a responsible data architecture.

Additionally, cost optimization, sustainability, and long-term scalability must guide architectural decisions as organizations move from pilots to production. Organizations must treat data as a strategic asset, build strong governance frameworks, and adopt a modern, interoperable architecture to confidently deploy Generative AI solutions that are accurate, secure, and ready to scale.

CHAPTER 8

Building a High-Performance Team for Generative AI

Introduction

Generative AI technologies have been evolving at a pace faster than ever anticipated. Since the launch of ChatGPT in November 2022 by OpenAI, the market has seen fierce competition from numerous players. The following year, 2023, saw the launch of Claude's long-context reasoning from Anthropic and the modality revolution with the release of multimodal capabilities from Gemini. The competition and race for innovation intensified with the release of several open-source models, including Meta's LLaMA, Mistral, and Stability AI, which marked the beginning of the accessibility revolution regarding who controls the Generative AI technology. Since then, new models have been launched almost weekly, benchmarks have shifted overnight, and what felt like state-of-the-art yesterday is already becoming a standard.

Every organization strives to adopt continuously evolving models to stay ahead in the competition. However, the real differentiator is not the model, but the team behind it. High-quality data, versatile models, fast

B. De, *Generative AI for Business Innovation*, https://doi.org/10.1007/979-8-8688-2682-5_8

vector databases, and scalable infrastructure are only enablers for building a Generative AI pilot. The real value of Generative AI investments lies only in a cross-functional, high-performing team that can combine all these elements to create an ethical and scalable solution. A team that can think across disciplines, experiment fast with all the evolving Generative AI technologies, and iterate safely to align with business values quickly can only deliver a product. A Generative AI initiative must start with a capable team built on a culture of innovation to take pilots to production. Visionary leaders and pioneering thinkers are essential to scale the Generative AI solution.

Generative AI is changing and will continue to change the way we work and the skills needed. C-level executives must therefore adapt fast to take a strategic approach focused on skill development and foster a culture of innovation. In this chapter, we will examine the special skills, organizational roles, culture, and approach required to build a winning team for an organization's successful Generative AI initiatives.

Why Teams Matter More Than Models

Generative AI models play a crucial role, but ultimately, it is a skilled and collaborative team that drives success. People with the proper knowledge and skills can enable effective integration and application of AI tools, manage complex data, evaluate outputs, provide feedback, and adapt to rapidly changing technological landscapes. Teams can drive better results by combining the right mix of tools, leveraging human and AI collaboration for content creation, and proactively addressing ethical and security challenges that AI alone cannot resolve. The following are some of the areas where a Team plays a vital role for Generative AI projects to ensure success, scalability, ethics, and real business impact:

- Business Alignment and Value Realization

- Ethics, Security, and Compliance

- User Experience and Accessibility

- Navigating Data Challenges

- Model Development and Refinement

- Integration and Deployment of Models

Key Takeaway: Tech is a commodity. People, culture, and governance are the differentiators. Hence, your investment in talent and culture determines ROI more than your choice of LLM provider.

Strategic Team Models for Generative AI

There is no one-size-fits-all. CTOs/CIOs must select an operating model that aligns with their enterprise maturity and scale:

- **Centralized Center of Excellence (CoE)**

  - Central pool of AI talent shared across business units

- **Federated/Embedded Model**

  - AI experts embedded into product squads with governance support

- **Hybrid Model**

  - CoE for standards + embedded talent for execution

Table 8-1 provides a decision lens for CTOs and CIOs to select the right model based on enterprise maturity, scale, and budget.

Table 8-1. *Decision Lens to Select the Right Operating Model*

Features	Centralized center of excellence (CoE)	Federated/ embedded model	Hybrid model
Core philosophy	Center of expertise.	Speed and context.	Governed empowerment.
How does it work?	Business units present needs to the central CoE, which prioritizes projects, develops solutions, and delivers or supports them as services.	AI talent collaborates daily within cross-functional product teams alongside product managers, designers, and engineers to address specific business challenges.	A central CoE handles strategy, governance, platform management, tooling, education, and standards, while embedded AI practitioners focus on innovation, execution, and daily development within business units.
Where AI talents reside?	Part of a single central team.	Distributed within business units/ product teams.	Embedded in both the core CoE and business units.
Advantages	Consistency, standardization, and deep expertise consolidation.	Speed, business relevance, and ownership.	Balances innovation with governance; most scalable.
Challenges	Can become a bottleneck; slower; may lose business context	Risk of duplication, inconsistent standards, and siloed expertise.	Complex to manage; potential for tension between central and embedded teams.

(continued)

Table 8-1. (*continued*)

Features	Centralized center of excellence (CoE)	Federated/ embedded model	Hybrid model
Ideal use case	A financial services or healthcare company in the early stages of Gen AI adoption, where managing risk and ensuring compliance is paramount.	A tech-first company, like a large SaaS provider or digital native business, where product teams are empowered and need to move extremely fast.	Almost any large enterprise (e.g., Fortune 500 retailers, manufacturers, banks) that is moving beyond initial pilots and needs to scale Gen AI responsibly across multiple divisions.

Developing the Skills for Generative AI

Building innovative, Generative AI solutions and scaling them from pilot to production requires a combination of diverse skill sets, both technical and interdisciplinary. Across the organization—from senior leaders to consultants and developers—baseline AI literacy should be established, with advanced proficiency concentrated in specialized roles. Organizations must invest in training their workforce on Generative AI models and transformer technologies, neural architectures, prompt engineering, model training and fine-tuning, vector databases, LLMOps/ GenAIOps, and high-performance computing.

Following are the broad category of skills needed for successful Generative AI adoption:

Foundational AI and Machine Learning skills to understand how Generative AI models work.

This includes the following:

- Deep understanding of neural architectures and various foundational transformer models, diffusion models, and other architectures like Variational Autoencoders (VAEs) or Generative Adversarial Networks (GANs)

- Model training and fine-tuning to adapt base models to specific tasks and domains

- Prompt engineering to systematically design and test prompts that reliably elicit designed behavior from foundational models

Data Engineering and Management skills to ensure high-quality data for model training and fine-tuning. This includes the following:

- Building and managing data pipelines to collect, clean, deduplicate, and preprocess massive datasets for training and fine-tuning

- Data labeling and curation for training and fine-tuning models efficiently, and improving the reliability of the models

- Vector database management to store, index, and retrieve high-dimensional vector embeddings needed for the RAG application

- Data governance and provenance to track data lineage and address data privacy

Software Engineering and MLOps skills to build and deploy reliable, scalable, and maintainable Generative AI solutions. This includes the following:

- Mastery of traditional software development practices using languages like Java, Python, Go, etc.

- Skills for automating and managing the end-to-end ML life cycle that includes

    - Automating the build, test, and deployment of fine-tuned models

    - Managing complex training and inference pipelines using tools like Airflow

    - Tracking model versions, parameters, and experiments using platforms like MLflow or Weights & Biases

    - Packaging and deploying containerized models

- Deep knowledge of cloud providers and services provided by them for distributed computing to parallelize training and inference across multiple GPUs/TPUs and scaling applications.

Infrastructure and Deployment skills are required to deploy and run LLMs cost-effectively. This includes the following:

- Understanding of high-performance computing, like GPU/TPU clusters and high-speed networking

- Quantization, pruning, and model distillation skills to optimize the model performance

- Experience with inference services like Triton Inference Server, TensorRT, or vLLM to achieve high throughput and low latency while serving models

Product Design and Human-AI Interaction skills to build AI systems that humans can easily use and trust. This includes the following:

- Designing UX for AIs that demonstrate the transparency of AI systems with things like confidence score, source citation, and explanation for outputs

- Allowing users to guide and refine outputs

- Building trust and safety in the application by gracefully handling hallucination and errors from AI models

Safe, Secure, and Responsible Production Deployments of Generative AI models and applications to ensure that the application generates safe and fair outputs. This includes the following:

- Developing mitigation strategies and content filtering systems to ensure that the AI models do not generate any toxic, harmful, or biased content

- Protecting AI models from jailbreak or deliberate adversarial attacks

- Understanding of the evolving AI regulatory landscape (e.g., EU AI Act) and data privacy regulations (GDPR, DPDP, CCPA), and copyright laws to ensure that the Generative AI outputs comply with the law of the land

All these skills are essential to move Generative AI solutions from pilot to full-scale production. Missing any of them will limit the performance, safety, and scalability of the application.

The Full-Stack Generative AI Team: Critical Roles and Evolving Skill Pillars

Generative AI has fundamentally re-shaped how software is built today, from design to deployment. Generative AI can suggest a robust and scalable design that meets business requirements. Tools like GitHub's Copilot, GitLab's Duo, Codeium, and CodeWhisperer are widely used for real-time code generation, debugging, and fixing issues. They have significantly improved overall development productivity and accelerated the pace of innovation.

Generative AI is not just about using tools; it is about integrating Generative AI APIs into software to create powerful and innovative functionalities. This requires specialized knowledge and skills along with high rigor, structure, and attention to detail. AI Engineers should demonstrate strong ownership, systems thinking, sound judgment, disciplined experimentation, critical thinking, and advanced problem-solving abilities. These qualities, along with clear communication, make them more effective in the fast-changing Generative AI world. To succeed, organizations need cross-functional teams that can work with a diverse set of technologies and move fast without breaking things. Such teams are built by combining several complementary roles as follows:

- **Strategic and Leadership Roles**

  These roles ensure that the Generative AI initiative aligns with the business goals and operates responsibly.

  - **Executive Sponsor/Head of AI**: The Head of AI will champion the Generative AI initiative at the highest level. They will ensure that the Generative AI initiatives are aligned with the overall business objective and are responsible for securing strategic funding. With strategic vision, business acumen, and budget oversight, they are ultimately accountable for the ROI. Strong executive leadership is essential to ensure a high probability of success for Generative AI initiatives.

- **Product Manager:** A product manager translates business problems into technical requirements. They define product vision, prioritize features, manage backlogs, and own the experience of the AI product. With a deep understanding of AI capabilities and limitations, as well as user-centric design and roadmap planning, they bridge LLM capabilities with real-world use cases. They can guide when to use or not to use RAG for building the use case.

- **AI Ethicist/Responsible AI Lead:** An AI Ethicist is responsible for establishing guidelines for safe, fair, ethical, and transparent use of AI systems. They conduct audits, establish guardrails, and ensure that the application complies with both internal ethics policies and external regulations. With a background in philosophy and knowledge of law and sociology, they monitor and ensure that the application is safe for use in society, while also adhering to regulatory compliance requirements.

- **Technical and Research Roles**

These roles involve conducting research, building, and deploying Generative AI models and their applications.

- **Data Engineer:** A Data engineer builds the data pipelines that prepare data for AI systems. They ensure that the ingested data is clean and of high quality through cleaning, validation, governance, and lineage tracking. They transform often messy and unstructured data into a labeled, contextual, and embedding-ready format for AI model training and inference.

- **ML Engineer:** A Machine Learning (ML) Engineer is responsible for the end-to-end ML life cycle, encompassing data preprocessing, model training, evaluation, optimization, deployment, and monitoring. Large Language Models like GPT-4, or open-source alternatives like Llama, come pre-trained with general knowledge of the world and language. Hence, ML engineers would focus on fine-tuning the Large Language models, if required, and refining them through experiments and live feedback loops to build robust, scalable, and efficient systems.

- **Research Scientist**: A research scientist possesses deep, specialized skills in advanced AI/ML technologies, often with doctoral degrees, who focus on exploring and experimenting with the latest developments in transformer architectures and fine-tuning state-of-the-art foundation models. They aim to solve complex problems for advanced product development.

- **LLMOps Specialist**: They specialize in industrializing AI by automating training, managing CI/CD, versioning prompts, and monitoring everything in production, from latency to content safety. They also monitor the hallucination rates for the models.

- **Infrastructure Engineer:** The infrastructure engineer is primarily responsible for setting up the cloud or on-premises infrastructure to deploy containerized LLM models and AI applications. They specialize in scaling GPU workloads while keeping spend within budget. They are also responsible for setting up distributed vector databases and serverless orchestration.

- **AI Engineer:** In an enterprise, an AI engineer focuses on **productionizing AI**. They are responsible for designing, building, deploying, and scaling enterprise-grade AI-powered systems to solve real-world problems. They ensure reliability, security, and scalability of the solution. An AI Engineer plays a very strategic role and sits at the intersection of data, models, cloud, and application engineering. They would be responsible for building and fine-tuning ML/LLM models, implementing RAG pipelines, optimizing model performance (cost, accuracy, and latency). They are also responsible for designing data pipelines for AI model training and inference, ensuring data quality and governance standards, and ensuring those systems are monitored, transparent, and built with responsible AI principles in mind.

- **Design and Experience Roles**

  These roles design the user experience for the application, ensuring it is safe, easy to use, and trustworthy.

- **UX Designers with AI Specialization:** They design user interactions for AI systems to enhance human-machine collaboration and drive adoption. They focus on transparency, fairness, trust, and access control through a human-centered design approach. They ensure that the application is reliable, safe, and build user confidence with better prompts. They also design error handling to manage issues like hallucinations gracefully, protecting users from risks.

- **Prompt Engineer:** A prompt engineer is a new role in the field of Generative AI. They design user inputs that the AI system can understand to produce the desired responses. Prompt engineers design, test, and optimize text prompts to interact with foundational models. They should have excellent written communication skills and a willingness to experiment with. Overall, prompt engineers blend technical, analytical, and linguistic skills to enhance AI system interactions and build a reliable and user-friendly Generative AI application.

- **Operational and Governance Roles**

These roles ensure that the application continues to operate effectively, safely, and reliably.

- **AI Security Specialist**: With deep expertise in cybersecurity, an AI security specialist focuses on mitigating security threats that are unique to AI applications, such as prompt injection attacks, data poisoning, and model theft.

- **Legal and Compliance Counsel:** They play an active role in addressing legal, regulatory, and ethical matters. They would advise on the intellectual property of Generative AI outputs, data privacy laws, and terms of service for third-party models whenever consulted.

Innovation Culture and Mindset Is the Differentiator

For any CTO or CIO, one of the biggest challenges in building Generative AI solutions is not the technology itself but creating the right mindset and team culture. More than deploying an AI model, the real challenge lies in building a culture that sustains momentum, manages uncertainty, and drives continuous improvement. It is the collaborative culture that drives execution.

Facilitate Collaborations and Communication

Everyone must think beyond their boundaries in terms of the systems that make up the entire application for better collaboration. Prompt engineers must not only think about how their prompts will elicit the best response but also how they impact the overall latency of the application. Team members building the Generative AI application must connect the dots across data pipelines, infrastructure setup, model performance, UX decisions, and risks involved to make real, lasting value. Below are some of the best practices for cross-functional coordination between teams to

- Ensure **collaboration among all teams through regular communication**. This helps align AI development with business needs.

- Implement **human oversight and governance** to validate the quality of AI output and consider the ethical implications.

- Conduct **ongoing upskilling and knowledge-sharing** sessions to keep the team informed on Generative AI advancements and challenges.

Build a Culture of Openness and Learning

An organization must build a culture that accepts failures and learns from them. Executives can promote this culture by sponsoring Generative AI education for their workforce. People at all levels must develop familiarity with Generative AI. Team members should also feel empowered to speak up early and question assumptions openly. Such a culture of intellectual honesty not only avoids costly mistakes but also drives better outcomes.

Experiment, and Iterate Fast

Getting everything right the first time may not always be practically possible, mainly due to the limitations of the evolving technology landscape. Hence, experimenting, learning, and iterating quickly should be the strategy to adopt the evolving AI technologies. Build a framework that provides quick feedback. The system should be designed to support continuous validation, with fast and safe rollback when needed.

Drive Success Through Shared Ownership

A culture of shared ownership makes everyone feel responsible for success and failure. This improves cross-functional collaboration and velocity. In the long run, it helps to improve resilience and scalability.

Challenge the Status Quo to Unlock Innovation

Curiosity, continuous learning, and questioning the status quo enable fearless exploration. This is important in a continuously evolving technology landscape of Generative AI. It helps the team explore new technology options, learn fast, and build together.

Technology alone doesn't determine the success of a Generative AI program—culture does. A culture that encourages learning, openness, and collaboration becomes the true differentiator in sustaining and scaling AI initiatives. Organizations that learn faster than the pace of AI innovation will ultimately outperform those that simply adopt new AI tools.

CIO/CTO's Playbook for Talent Strategy

This playbook will help CTOs and CIOs define and create the right talent strategy for their organization's Generative AI initiative. It is structured in five phases, namely, Define, Build, Operate, Scale, and Evolve. The talent strategy should also evolve with organizational AI Maturity. What works for experimentation often differs from what is needed to scale enterprise-wide.

Phase 1: Define the Talent Strategy and Business Alignment

- **Align the Talent Strategy to Business Objectives**: Identify and prioritize the portfolio of use cases to be implemented using Generative AI that will solve the business problem. Let the validated use cases define the hiring needs rather than starting with predefined roles like "LLM expert."

- **Define the Operating Model Blueprint That Outlines Where Talent Will Be Situated:** Will they be part of the Center of Excellence (CoE), embedded within business units, or deployed via a hybrid structure? Specify funding ownership, decision rights, and governance mechanisms for talent allocation, and clarify how teams will collaborate across organizational units. This will determine how and who will sponsor the required talents, as well as how they will coordinate with the rest of the organizational units.

- **Map the Core Competencies Required for Your Prioritized Use Cases:** This will help to define the key roles needed and the skill matrix for the team.

Phase 2: Build the Team (Acquisition and Development)

- **Build vs. Buy vs. Partner to Acquire the Right Skills:** Building and upskilling existing talent is the most powerful lever for growth. Invest in a structured training program to upskill the best in-house engineers, data scientists, and product managers keen to learn and work on AI solutions. Having a community of practice for Generative AI can foster a culture of learning. As a second approach, acquire external talent with strong foundational ML skills who have prior experience in building Generative AI solutions deployed in production. As an alternative, use strategic partners for initial acceleration, proof-of-concepts, and to fill specific skill gaps.

- **Sell Your Vision to Solve Interesting Problems with Generative AI Technology:** Top AI talent seeks to work on challenging and interesting problems.

Phase 3: Operate for Impact (Execution and Enablement)

- **Start with a Small Cross-Functional Team:** This team, with its various roles, should be tasked with solving and implementing the initial, prioritized use cases. It helps to establish the first set of internal Generative AI experts.

- **Provide Guardrails** with clear policies on data privacy, model usage, and production deployment that the implementation team must follow. The central CoE team must provide a curated, secure, and approved stack for Generative AI implementations for enterprise-wide adoption. This makes it easy for teams to "do the right thing" and hard to make catastrophic mistakes.

Phase 4: Scale Responsibly (Governance and Democratization)

- **Formalize the Hybrid Model:** Form the CoE team that will enable other product teams or business units across the enterprise. They will define the standards and implementation best practices, manage the AI platform and infrastructure, and provide training and consultations for other embedded teams.

- **Embed and Grow Talents Within Business Units:** Grow the AI expertise within the product teams and business units with new hires and internal training.

- **Institute Robust Governance:** Establish a cross-functional AI Governance Council to review high-risk use cases *before* they are built. Reduce, detect, and contain hallucinations, bias, toxicity, and data leakage, while also managing costs via evaluation, guardrails, and human-in-the-loop workflows.

Phase 5: Evolve the Capability (Culture and Continuous Learning)

- **Define a Career Ladder for Generative AI specialists:** This will increase talent retention.

- **Measure Talent KPIs:** Measure the effectiveness of building and elevating AI skills and the *ability to attract and retain the critical talent needed for long-term success* within the organization. Also, track and celebrate the productivity of AI teams and their innovations.

- **Stay Ahead of the Curve:** Allocate time and budget for the top talent to research new models and techniques.

Facilitate participation in conferences and encourage experimentation without fear of failure and move forward.

Executive Checklist: Five Questions that Every CTO/CIO Should Ask

- Do we have a clearly defined **AI operating model** (CoE, embedded, or hybrid) with a named owner, documented decision rights, and a review cadence?

- Do we have a funded **upskilling and cross-training plan** for employees with accountable owners, measurable skill benchmarks, and quarterly progress tracking?

- Do Generative AI initiatives have assigned business owners, defined **ROI and adoption KPIs**, and a reporting cadence that is reviewed at leadership level?

- Is there an **operationalized AI governance framework** with designated stewards, compliance metrics, and scheduled audits?

- Do we track learning and innovation through defined metrics (e.g., experiments run, ideas scaled, reuse rate), with executive ownership and periodic review, to foster a **culture of learning and innovation**?

Future Outlook: The Generative AI Team of Tomorrow

The Generative AI team for tomorrow is going to be markedly different than what it is today. The rise of agentic AI, accompanied by a shift from LLMs to SLMs tailored for specific domains, is expected to drive a shift in skill sets and the emergence of new roles. A greater emphasis will be placed on human-AI collaboration. AI will be seamlessly integrated into tools and workflows to drive real business value. As a result, current roles will evolve, needing enhanced critical thinking, increased adaptability, and learning agility.

With AI handling more low-value, repetitive tasks, human team members will be able to focus more on higher-value work that requires creativity, strategic thinking, and emotional intelligence. The focus of the future team will shift from technology for building and fine-tuning models to the strategic and ethical application of AI in solving real-world problems. The CIO/CTO role will evolve from technology leader into an **orchestrator of hybrid workforce consisting of humans, AI agents, and autonomous systems**.

Conclusion

Unlocking the transformative potential of Generative AI needs a collaborative culture of innovation. With the rapid pace of evolving technologies, as new models, tools, and frameworks are released every month, you need people who can thrive in ambiguity and continually master new skills. **People and culture are enduring assets**. A high-performance team must blend technical excellence, governance, product thinking, and collaborative culture.

Generative AI teams will continue to evolve in tandem with these technological shifts. To stay ahead, CTOs and CIOs must invest in a comprehensive approach for skill development and encourage continuous learning and experimentation. Leaders can set their Generative AI team apart by embracing flexibility, encouraging growth, and fostering a strong learning culture, and establishing clear performance metrics and accountability mechanisms to track outcomes and impact.

CHAPTER 9

Scaling Generative AI: From Pilot to Enterprise Value

Introduction

Generative AI holds an excellent promise for revenue growth and productivity improvements. However, Gartner forecasts that up to 85% of AI projects will fail. According to the MIT report "State of AI in Business 2025," 95% of Generative AI projects in companies are failing to deliver measurable business impact or revenue growth. The real reason is not the models, and there is no single reason for these failures. Unclear objectives, an incorrect choice of use case, insufficient data readiness, a lack of in-house expertise, organizational culture, leadership intent, and a skewed budget for shiny use cases are among the main contributing factors.

CIOs are taking a PoC approach to test AI strategies and technologies, keeping their stakes low in case of failure, without full commitment. PoCs for Generative AI are being approved more easily, mainly due to pressure from boards and CEOs to deliver faster outcomes driven by AI marketing campaigns. Lots of PoCs are underway to test different ideas and experiment with Generative AI technologies at a small scale. Once the technical feasibility is established, a pilot implementation gets approved.

© Brajesh De 2026

B. De, *Generative AI for Business Innovation*, https://doi.org/10.1007/979-8-8688-2682-5_9

Pilots have become the easiest way for CXOs to signal the adoption and progress with Generative AI. But most of the pilots do not make it to wide-scale production deployment as they fail to overcome the challenges of unclear ROI, insufficient data readiness for AI, and a lack of in-house skills.

It needs structure, governance, and precise business alignment to see the success of pilot implementation making it to production. Successful Generative AI adoption requires a well-designed, well-prioritized use case, a strong data foundation, ethical guardrails, and cross-functional governance. The business leaders must focus on measurable ROI, operational impact, organizational readiness, and long-term scalability. A successful production rollout of a Generative AI pilot requires adherence to best practices for Generative AI implementations.

In this chapter, we will cover critical, non-negotiable factors that can determine the success of a Generative AI project. The best practices for each of these strategic considerations is discussed in this chapter.

From Pilot to Production: The Seven Non-Negotiables

The following are seven non-negotiable factors that can determine the success of a Generative AI initiative—moving from pilot to production.

- Business-aligned use case

- Production-grade data foundation

- Modular and scalable architecture

- Governed LLMOps

- Responsible AI and security by design

- Organizational readiness

- Measurable ROI

Strategy and Planning

Use Case Selection

PoCs and pilots are often the easiest way for technologists to demonstrate the technical feasibility of AI and build familiarity with Generative AI technologies. In reality, the use cases for these PoCs and Pilots primarily focus on the wow factor rather than real value. They are often not aligned with the primary business objectives. This is where the gap starts. The use case selected for the initial PoCs and pilots must align with the business goals and aim to solve a real problem the business is struggling with. If Generative AI can solve that problem, the pilot use case will be poised for successful adoption in the mainstream. From a business perspective, the use case must focus on accelerating revenue, improving operational efficiency, driving customer delight, or differentiating the product. The use case and solution must explain how they will fit into the workflow and integrate with other systems. This will drive the adoption of the Generative AI solution faster.

What looks like a feasible use case in the PoC phase may fail in the pilot phase as it scales and attempts to meet performance needs. It is important to fail fast and identify challenges early when choosing the proper use case. The feasibility assessment of the use case must consider the availability of the required data, the complexity of integration with other systems, production workflows, change management efforts, compliance risks, and the users who will ultimately consume it. Real users and real conditions must be considered while selecting the use case. Without that, pilot experiments are very likely to fail.

Buy, Boost, or Build

Many organizations invest in building in-house Generative AI solutions rather than purchasing AI tools from specialized vendors. Building and training Generative AI models can be significantly more expensive than buying readily available AI models. Even building a proprietary AI tool in-house can incur high costs, lowering ROI compared to purchasing an AI tool from a specialized vendor. In many cases, customizing or enhancing a purchased Generative AI solution can deliver results faster.

Today, many off-the-shelf Generative AI tools/solutions perform exceptionally well across general tasks such as coding, text processing, image generation, and standard chatbots. Buying a ready-made AI solution gives a quick start without having to invest in development and fine-tuning. This reduces release cycles and provides a predictable subscription cost. Even updates are managed by vendors, reducing software management overheads. However, they offer limited customization, carry risks of vendor lock-in, and provide no competitive advantage.

Enterprise can enhance or boost AI tool/solution capabilities by integrating proprietary and sensitive data to build use cases that rely on contextual, enterprise-specific data. Boosting a solution can be achieved by fine-tuning a model to perform better in a specific setting, or by using retrieval-augmented generation, which involves feeding company-specific information into the Generative AI model to improve its accuracy. This provides better accuracy for specialized tasks and reduces hallucinations without the full cost of building. However, this adds to the operational/usage expenses and requires data governance practices.

Building a fully custom AI solution with proprietary data can be the only option that demands stringent regulatory compliance, higher security, and total control over IP. Building proprietary Generative AI solutions requires significant additional investments in infrastructure, data, and talent, which can drive costs.

Organizations need to understand their current strengths and weaknesses and consider the following factors to decide whether to build, boost, or buy a Generative AI solution. Table 9-1 provides a decision framework to help organizations determine when to build, boost or buy a Generative AI solution.

Table 9-1. *Framework for Build, Boost, and Buy Decision*

Factors	Build	Boost	Buy
Time to market	A custom solution can take from months to years	Provide a moderate time-to-market	Accelerates time to market within weeks
Customization and control	Provides full control that is tailored to business needs	Moderate control using proprietary data and fine-tuning	Limited control; designed for general purpose
Competitive advantage	Gives a significant competitive advantage, powered by proprietary data and assets	Provides relevant and accurate results with proprietary data	Can limit the differentiation and introduce vendor lock-in
Data privacy/ security	Offers complete control on data handling and compliance	Requires robust data governance and security processes	Limited control, relying primarily on the vendor's compliance and security measures
Cost	High up-front investment, lower long-term usage cost	Moderate initial and ongoing costs	Subscription/ Usage-based pricing provides a lower initial cost. But overall long-term cost can be high
Maintenance/ updates	Internal responsibility. Updates have to be managed continuously	Vendor handles core model updates; internal team manages enhancements	Managed by vendor

To summarize, the following decision heuristics will be helpful:

- **Build**: If regulatory risk is high

- **Buy**: If differentiation is low

- **Boost**: If proprietary data is core

Cross-Functional Team

A cross-functional team is not optional. It is critical to roll out Generative AI solutions to production successfully.

Implementation of Generative AI applications cuts across data infrastructure, UI/UX, business process workflows, security and ethics. Hence, a cross-functional team is required to address different aspects. A shared ownership between IT, security, legal, compliance, and business is needed to address risks of Generative AI solutions due to hallucination, data leakage, model drift, performance, uncontrolled inference costs, and regulatory and ethical exposure. A collaboration between business, domain, and UX roles is needed to take cool demos to production that demonstrate measurable business outcomes.

A collaborative, cross-functional team for Generative AI **reduces time-to-value**. It provides a **faster, safer path to productio**n, with fewer reworks post-pilot and early identification of security and compliance gaps. An early collaboration with business teams **builds trust and drives adoption** of Generative AI solutions with clear accountability. It also helps to **control costs and build a sustainable solution** by selecting the right model for the right task, putting budget guardrails and monitoring usage and cost. By checking for bias and safety, the legal and ethics team can help to make the Generative AI solution more responsible and accountable for their outcomes. This **reduces regulatory and reputational risk** when the solution is rolled out to production for all.

Table 9-2 shows the key roles and their primary responsibility in a cross-functional Generative AI team.

Table 9-2. *Key Roles and Responsibilities in Cross-Functional Generative AI Team*

Roles/Function	Primary responsibility for Generative AI solutions
Business/domain owners	• Identify and prioritize high-impact use cases aligned to business goals. • Define success metrics, KPIs tied to business outcome.
Data engineers	• Develop and manage data architecture. • Build scalable data pipelines and ETL processes. • Ensure data cleanliness, accuracy, and reliability. • Create embeddings and store data in vector stores.
AI/ML engineers	• Select efficient and cost-effective LLM/ SLMs. • Deploy models into production environments. • Design and build RAG solutions. • Train and fine-tune models and evaluate outcomes.
Cloud/platform engineers	• Build and manage a scalable MLOps infrastructure. • Architect the underlying compute, networking, and storage for high-throughput, low-latency inference. • Implement robust monitoring, fault tolerance, and auto-scaling for enhanced reliability of cloud platform. • Optimize costs of GPU/accelerator usage.
Security and compliance team	• Embed security controls for identity and access controls, encryption and network isolation. • Ensure compliance with data privacy regulations, access controls, and regulatory compliance. • Establish governance for model access, audit trails, and responsible AI principles.

(continued)

Table 9-2. (*continued*)

Roles/Function	Primary responsibility for Generative AI solutions
Legal team	• Navigate Intellectual Property (IP) and assess risks for training data and model outputs. • Ensure regulatory compliance and accountability for model behaviors and errors. • Implement governance for transparency and bias mitigation. • Manage contractual and liability exposure.
UX/product designers	• Design intuitive interfaces that manage probabilistic, non-deterministic outputs, including clear input guidance. • Build Trust, and explain AI capabilities and limitations transparently. • Design for human feedback loops in the flows. • Design AI solutions that can co-create with humans with iterative refinements.
MLOps/AIOps team	• Build end-to-end automated ML pipelines for data, model training, fine-tuning, evaluation, and deployment. • Design and manage serving infrastructure for high-throughput, low-latency inference. • Implement monitoring for model performance, data drift, system health, and business metrics. • Ensure model governance, security, and life cycle management and ensure compliance.

(continued)

Table 9-2. (*continued*)

Roles/Function	Primary responsibility for Generative AI solutions
Change management team	• Drive organizational adoption and mitigate resistance. • Work with leadership and teams to redesign business processes and clarify new roles and responsibilities. • Track adoption metrics, address barriers to adoption. • Lead cultural shift toward experimentation and drive responsible AI usage.

Table 9-3 provides an enterprise-grade RACI with decision and approval gates specifically designed to move Gen AI Pilots to Production.

Table 9-3. *Enterprise RACI with Approval Gates for Production Deployment of Generative AI*

	Gate 0 – Use Case Qualification	Gate 1 – Data and Architecture Readiness	Gate 2 – Pilot Validation	Gate 3 – Production Readiness Review	Gate 4 – Deployment Authorization	Gate 5 – Scale Post Launch Validation
Business/Domain Owners	A/R	C	A	A	A	A/R
Data Engineers	C	R	C	C	C	C
AI/ML Engineers	R	C	R	C	C	C
Cloud/Platform Engineers	C	A	C	R	R	R
Security & Compliance Team	C	R	C	R	C	C
Legal Team	C	C	C	R	C	C
UX/Product Designers	C	I	R	C	I	C
MLOps/AIOps Team	I	C	I	R	R	R
Change Management Team	I	I	I	C	R	R

A = Accountable (Decision Authority)   C = Consulted
R = Execution Owner   I = Informed

Foster an AI-Ready Culture

Taking pilots to large-scale production deployments requires a different mindset and thinking approach. An AI-Ready culture is often the missing link between a successful Generative AI pilot and a scalable, trusted production deployment.

Generative AI changes how people work with systems. It produces probabilistic outputs that need to be validated by humans, and outputs can evolve through iteration. Trusting AI outputs blindly can be very risky. On the contrary, low trust can become a significant barrier to the adoption of Generative AI solutions. Humans need to know how to use Generative AI applications to augment their decision-making process.

Pilots are typically developed by small, innovative, and highly motivated teams using controlled datasets. However, scaling the solution for production deployment would require redesigning the process and achieving broader user adoption with a significantly different cultural mindset.

Fear and skepticism about job displacement due to AI, misuse or lack of understanding on when to and when not to use AI are some of the other issues that can be addressed by building an AI-ready culture.

An AI-ready organization exhibits the following:

- Informed trust (not blind trust)

- Human-in-the-Loop mindset

- Responsible AI awareness

- Data and AI literacy across roles

- Continuous learning and experimentation

Organizations must do the following to build an AI-Ready culture that can help move Generative AI pilots to production.

Improve AI and Data Literacy

People within the organization must understand what Generative AI can and cannot do. They must be aware of how to deal with hallucinations, bias, and uncertainty in Generative AI. They must know how to validate the probabilistic outputs of Generative AI. AI literacy can drive responsible usage.

Build Ownership and Accountability Mindset

Generative AI should not be seen as a black-box. There must be clear answers to who owns AI outcomes, who validated them, and who is accountable for errors.

Accept Probabilistic System

The outputs of Generative AI can vary depending on the context and prompt. This is in sharp contrast to deterministic systems that consistently produce the same output. Production teams need to understand that and change their expectations. Confidence score matters, and it should be considered when validating or using Generative AI outputs.

Design Responsibly

An AI-Ready culture must inculcate a mindset and philosophy of designing Generative AI systems ethically and responsibly. Design should ensure that the solution is safe for mankind. Safeguards should be in place to prevent any misuse. Bias awareness should be built into the design. Ensure that data and application design promote fairness. Data privacy, security, and regulatory compliance must be baked into the design of the Generative AI system.

Encourage Experimentation with Guardrails

Innovation must continue after initial pilot implementations. Learnings from the pilots must be used to improve the solution going forward. The team must feel safe to iterate on prompts and workflows and fail fast within boundaries.

An AI-Ready culture comes with the following advantages:

- Higher adoption and business impact as AI becomes part of the daily work

- Faster transition from PoC to Production due to less resistance from operations, security, and legal teams

- Reduced operational and reputational risks with safe AI usage and better oversight that can detect and prevent early misuse

- Sustainable AI solution at scale as teams understand cost implications of prompts and models

Technology and Data

Modular and Loosely Coupled Architecture

Enterprise Generative AI are inherently multi-component—consisting of prompts, models, context retrieval, model inference, and fact verification. Scaling Generative AI solutions in production is a key requirement for their success. A modular architecture is needed to avoid a single point of failure, avoid vendor lock-in, scale each component independently, and manage costs. It provides the following advantages:

- Allows continuous experimentation without downtime

- Meets regulatory compliance through pluggable modules

- Adopt a multi-modal strategy to perform specific tasks based on their strengths

- Provides visibility into performance, quality, cost, and failures

Vendor lock-in to a single LLM provider can make switching prohibitively costly. Changing one component should not break or introduce risks to other components. Hence, the application must have a modular architecture built using an **API-First approach and microservices**.

Microservices can break down workflows like preprocessing, model training, and embedding generation into modular components. The core functional modules should include the following:

- **LLM Gateway**: Provides standardized API access to foundational models. It enforces authentication, rate limits, observability, and fallback routing.

- **Prompt and Policy Service**: Manages prompt templates, guardrails, versioning, and experiment tracking.

- **Orchestration Layer**: Coordinates multi-step reasoning flows, tool calls, and retrieval pipelines through workflow APIs.

- **Embedding Service**: Generates embeddings with batching and caching capabilities.

- **Vector Datastore**: Handles indexed retrieval with defined query contracts and metadata filtering.

- **Training and Evaluation Service**: Manages dataset ingestion, fine-tuning jobs, offline metrics, and model registry integration.

- **Application Logic**: Consumes AI services through stable APIs, remaining isolated from model-specific implementation details.

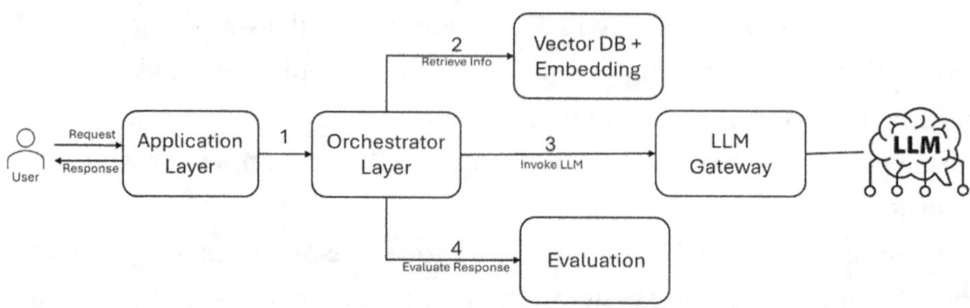

Figure 9-1. *Anatomy of an Enterprise Generative AI Solution*

Figure 9-1 shows the reference architecture of an enterprise Generative AI solution and the interaction between different components - the application, orchestrator, vector datastore, LLM and evaluation layer. All these components must communicate with each other using well-defined interfaces. REST APIs can be used for synchronous calls and event-driven approaches for asynchronous flows. Event-driven pattern using topics and queues enable loose coupling and horizontal scalability that may be required for heavy AI workloads such as training, batch embedding generation, and large-context inference.

Data Readiness and Governance

If data is not production-ready, Generative AI will never be production-ready. High-quality, traceable, accessible, and governed data is paramount for reliable Generative AI. Conduct a data audit to know the following before getting started with pilot:

- What data is available?

- Where is it stored?

- How is it structured?

- How clean is it?

- How easy is it to integrate and consume?

The following are some of the best practices to ensure data readiness:

- **Treat Data As a Product**

  - Ensure that the data is reliable to scale.

  - Every domain must have a defined owner and stewards for its data.

  - Data should have SLAs for its freshness, completeness, and accuracy.

  - Data contracts with structure, semantics, quality, and governance rules should be defined and published between consumers and providers.

- **Establish a Precise Data Classification and Sensitivity Model for Compliance**

  - Data must be classified up front, and its sensitivity should be defined.

  - Define AI-usage policies for the data to identify what can be used for training, embedding, and inference.

  - Mask or tokenize sensitive fields in the data before vectorizing.

  - Set up controls for privacy and role-based access.

- **Build Scalable and Auditable Data Pipelines That Go Beyond Pilot**

  - Design data pipelines for ingestion, validation, and quality check.

  - Pipelines must support incremental loads and schema evolution.

- Design separate pipelines to process training, inference, and observability data.

- Store raw, curated, and enriched data separately.

- Use Vector DBs that scale independently of compute to store enterprise data for RAG.

- **Automate Data Quality**

  - Define data quality metrics to be monitored like accuracy, completeness, timeliness, freshness, consistency, duplicate rates, etc.

  - Automate and ensure the highest levels of data quality checks.

  - Build pipelines that fail when quality thresholds are breached.

  - Monitor data drift.

- **Track Metadata and Lineage for Explainability**

  - Track data lineage from the source system to embeddings till response.

  - Maintain metadata for source system, response time, and transformation logic.

  - Store references returned by RAG.

- **Set Up Governance for Vector Data**

  - Track the lineage and provenance of vector data for compliance and address hallucination: *Source Data* ➤ *Chunking Strategy* ➤ *Embedding Model* ➤ *Index Version* ➤ *Usage in Retrieval.*

- Define and measure the quality of your vector embeddings and the retrieval process for high-quality retrieval and LLM response.

- Set up access control with granular permissions at vector/index level by integrating with IAM.

- **Separate Training Data from Inference Data**

  - Maintain separation between training, evaluation, and inference data.

  - Apply consent and anonymization rules.

  - Maintain dataset versioning.

- **Implement Robust Access Control and Least Privilege for AI**

  - Set up role-based access control at granular levels by integrating with IAM to control ingestion, embedding, and query capabilities.

  - Set up row-level and column-level security.

  - Set up secure endpoints for vector search.

- **Set Up Data and AI Observability for Continuous Monitoring**

  - Monitor data freshness, retrieval accuracy, and drift in embedding.

  - Capture user feedback, ratings and corrections.

  - Track hallucinations linked to data gaps.

  - Set up alerts for stale indexes and retrieval failure.

- **Set Up Human-in-the-Loop and Approval Workflow**

  - Manually validate for new data sources and high-risk domains.

  - Set up approval gates for data onboarding.

  - Enforce and conduct periodic governance review.

Invest in Scalable Infrastructure

Applications need to scale seamlessly in production. Otherwise, the transition from a pilot to production is very likely to fail. Hence, even the infrastructure for pilot implementation should be designed for production from day one. It is always advisable to use a scalable cloud infrastructure and containerize applications from the start. The following are some of the best practices to ensure that the infrastructure scales easily for production:

- Understand the NFRs for production early and set up a **production-like environment** for pilot implementation using cloud services, IAM, and networking. This will reduce the need for re-architecting to scale.

- Prefer **managed services** over self-managed for Kubernetes/App Services, databases, and vector DBs.

- Use **horizontal auto–scaling** for APIs and microservices.

- Use queue-based scaling for async workloads.

- Automate infrastructure setup using Infrastructure as Code for predictability and repeatability.

- Strive for **repeatability, predictable scaling,** and faster environment provisioning by automating infrastructure setup using **Infrastructure as Code.**

- Define a clear environment strategy with separate environments for Dev, QA, and Prod.

- Set up **automated CI/CD processes** for build, test, security scans, and deployment right from pilot implementations.

- Use **feature flags** to roll out features safely. Use Blue-Green or Canary deployments for production to ensure scalability.

Select the Right Models and Tools

Selecting the right model and tools is one of the most thought-through decisions for any Generative AI implementation. Making the correct choice during the pilot lays out the groundwork for a successful deployment in production. Many pilots fail because they lock into the wrong model or tool. The following are some of the best practices to be followed for selecting the right models and tools that will scale easily with production workloads.

- Select the model based on the use case for the Generative AI implementation.

- Evaluate models based on non-negotiable parameters like accuracy, latency, context window, cost per request, and regulatory constraints.

- Design for model portability to easily switch between models by abstracting models behind a model abstraction layer using frameworks like LiteLLM, LangChain.

- Design for a multi-model strategy with primary, fallback, and cost-optimized models.

- Start with smaller models that meet business and quality goals rather than bigger models. Light-weight SLMs can be more efficient and cost-effective for specialized workloads.

- Use pre-trained models with managed AI services that come with built-in scalability, reliability, and enterprise security, but plan for fine-tuning or custom builds for specific needs.

- Avoid the complexity of self-hosting models unless data sovereignty is non-negotiable.

A pilot becomes production-ready when models and tools are chosen for change—not perfection. Design the architecture using LLM gateway for a multi-model strategy so that teams can switch models as requirements evolve without rework.

Execution and Scaling

Start Small and Iterate Fast

This agile development philosophy drives innovation. Begin with a Minimal Viable Product (MVP) and iterate fast to reach production deployment in a controlled environment. The following are some of the best practices to be followed for this iterative approach:

- Start a Generative AI pilot with clear boundaries—a specific task, user group, or data domain.

- Agree on non-negotiable production requirements early in the first few sprints.

- In each development iteration, establish one specific "North Star" metric—such as accuracy, cost, or latency—to measure rigorously and guide all progress.

- Adopt a versioning strategy right from the start for code, prompts, configuration, and datasets.

- Iterate not just on code but also for the prompts and datasets.

- Ensure security is implemented and tested right from the first iteration. Every subsequent iteration must build on top of the previous one to add more security controls.

- Plan to implement failure modes for a graceful degradation as early as the 2nd or 3rd iterations.

- Include end-to-end logging, monitoring, and alerting as part of every sprint development.

Each iteration should produce a shippable increment that is marginally more secure, observable, and robust than the last.

Run Governed Pipelines for LLMOps

To ensure that Generative AI pilots successfully transition to production, governed pipelines for LLMOps must be set up from the start. Below are some of the best practices for the same:

- **Define Governance and Ownership**

  - Define a Lightweight LLM Governance Model.

  - Define all the leading roles for LLMOps along with the responsibilities and accountability.

  - Consider the following roles for LLMOps to start with—Prompt Owner, Model Owner, RAG/data owner, Risk reviewer, and Production approver.

- **Treat Prompts, Chains, and Agents as Code**

  - All prompts, templates, chains must be version-controlled in repository.

  - Use PR-based reviews and approvals for prompt changes.

  - Tag versions as *experimental, approved,* and *production.*

- **Govern RAG data sources from the start to reduce hallucinations**

  - Use curated and approved document sources for RAG.

  - Tag data with metadata information for owner, freshness, and sensitivity.

  - Use metadata and filters to refine filtering and exclude irrelevant or outdated chunks of data.

  - Version embeddings to enable rollback.

  - Calibrate with confidence thresholds to gracefully admit uncertainty.

- **Use Production-Grade CI/CD Pipelines in the Pilot**

  - Set up automated deployments for prompts, agents, and RAG indexes.

  - Use Infrastructure as Code to set up and configure separate environments for Dev/QA and Prod.

- **Build Automated Evaluation and Quality Gates**

  - Build a golden dataset to test diverse input–output pairs that represent real-world cases. Perform A/B testing of diverse prompts against the golden dataset.

  - Evaluate the groundedness/faithfulness of the output with respect to the retrieved context. Faithfulness score must be > 0.9.

  - Automate the evaluation of output for correctness, safety, and style using traditional metrics and using LLM-as-a-judge.

  - Use Guardrail libraries like Microsoft Guidance, NVIDIA NeMo Guardrails, or Azure AI Content Safety to enforce rules on topics, tone, and safety.

Prioritize Security and Compliance

Design AI components for security and safety from day one and not as an afterthought. Security cannot be retrofitted. It must be secure by design. Every AI component must have a threat model and undergo a pen-test, along with other security testing, early in the SDLC.

The following are some of the best practices and essential guardrails to bake into the security design:

- **Secure the Interaction with Models**

  - Implement defense against prompt injection by validating and sanitizing input prompts for malicious code or excessive length.

  - Screen responses for sensitive data leakage (PII, credentials, proprietary info) before display.

- Secure API endpoints for AI models using strong authentication (OAuth2, API keys), rate limiting, and DDoS protection.

- Implement access control using the principle of least privilege to give users/systems with minimal model access as required.

- Prevent excessive model use using rate limiting to throttle requests.

- Separate the production model from training/ experimentation using network isolation.

- **Ensure Data Security and Compliance**

  - Sanitize data for input/output security using input validation and filtering.

  - Mask PII data before processing using automated redaction tools.

  - Implement end-to-end encryption for data at rest using AES-256 or stronger algorithms.

  - Set up RBAC with the least privileges and network isolations to protect datasets.

  - Ensure compliance to regulatory (GDPR/CCPA/ DPDP) and domain requirements (HIPAA/PCI).

- **Build Auditability and Oversight into Operations**

  - Implement comprehensive logging and monitoring to log all prompts, responses, model versions, and user interactions.

  - Mandate a human review for high-risk decisions. Clear escalation and override procedures must be defined.

- Be prepared to explain how the AI response was generated. Ensure that explainability and transparency are built in for regulatory compliance, like GDPR's right to explain.

- Test for unintended bias in output during the pilot phase itself.

Enforce Responsible AI Checks

Lack of Responsible AI guardrails is one of the main reasons for the failure to move a Generative AI pilot to production. Human oversight at every stage of the AI life cycle (train, operate, and evaluate) to guide and validate the outcomes enhances the reliability, fairness, and decision quality. Organizations that bake in responsible AI into their SDLC for Generative AI move to production faster. The following principles should guide a responsible AI check:

- Fairness and Bias Mitigation

- Transparency and Explainability

- Privacy and Security

- Reliability and Safety

- Accountability and Human Oversight

The following are some of the best practices to ensure a responsible AI solution when it is deployed in production:

- **Embed Responsible AI into the Delivery Life Cycle (Shift Left)**

  - Define responsible AI acceptance criteria along with functional requirements.

- Include ethics, safety, and compliance checks as part of design review, model selection, and prompt and RAG architecture.

- **Establish Clear Ownership and Governance Early**

  - Create a lightweight governance council for responsible AI.

  - Define clear roles and owners for the model, data, technology, and compliance risks.

  - Define an escalation path for ethical and safety issues.

  - Define clear thresholds where human review and overrides are mandatory.

- **Conduct Data Quality**

  - Validate data provenance to ensure it is accurate, complete, and free from harmful historical biases.

  - Validate sensitive data exposure and regulatory compliance.

  - Mask, tokenize, or redact sensitive data before embedding or prompting.

- **Perform Security and Stress Testing**

  - Conduct comprehensive testing to ensure that the system is robust and reliable in production.

  - Ensure that the model and the solution pass the vulnerability assessment.

- **Conduct Bias, Safety, and Hallucination Testing**

  - Create a test suite for bias scenarios and integrate with CI/CD.

  - Measure hallucination rates, answer grounding, and citation accuracy (for RAG).

  - Implement guardrails to moderate toxicity in output.

- **Documentation and Transparency**

  - Document use-case intent.

  - Document details about model design, limitations, and known risks.

  - Document details about training data sources.

  - Integrate with explainability tools to ensure that the model's outputs can be interpreted and explained.

- **Monitor Continuously Post-Launch**

  - Establish automated monitoring systems to track model performance, data drift, and unexpected bias in real-time.

  - Maintain audit logs to track the AI's actions and decisions for future reviews or regulatory scrutiny.

- **Establish Human-in-the-Loop (HITL) for High-Risk Scenario**

  - Define clear thresholds where human review and override are mandatory.

  - Set up governance boards to monitor trends and check for bias and errors.

- Route uncertain AI outputs from complex transactions to human experts for review before action.

- Capture user corrections and feedback as learning signals.

Upskill to Address Talent Shortage

Upskilling the workforce on AI, including Generative AI, enables organizations to maximize their investments in technology and gain a competitive edge. The talent gap can slow down the adoption of Generative AI technologies within the company. The following are the best practices organizations need to adopt to ensure successful AI upskilling:

- **Strategic and Cultural Best Practices**

  - Assess current skills, forecast future needs, and identify specific Generative AI skill gaps.

  - Get leadership buy-in to champion AI trainings.

  - Foster a culture of continuous AI learning and adoption to grow talent internally.

  - Incentivize new AI learning through rewards and defined career paths.

  - Implement a mentorship program to guide newcomers in navigating the AI complexity.

- **Training and Skill Development**

  - Use AI for AI upskilling.

  - Define separate training tracks for each role, namely, business and product leaders, architects and tech leads, AI engineers, and data and platform team.

- Start with AI literacy and responsible AI foundations.

- Conduct hands-on and practical training for enterprise-grade AI, covering

  - Data preparation and governance

  - Model selection and evaluation

  - LLMOps (CI/CD, monitoring, rollback)

  - Cost optimization and performance tuning

- Measure skill outcomes for each role with competency outcomes and not the training hours. Use defined proficiency levels, hands-on assessments, and certification-style validations that demonstrate real delivery readiness.

Monitor and Measure Continuously (Real-Time Observability)

Continuously track metrics, retrain models, and adapt as

- **Implement Comprehensive, LLM-Specific Monitoring and Observability**

  - Log Everything—Prompt/response pairs, latency, token usage, costs, model used, user ID (anonymized).

  - Monitor key performance metrics in real time like latency (P50, P95), throughput, error rates.

  - Track token consumption per user/feature, cost trends.

- Track model and embedding drift.

- Set Alerts for latency spikes, increased error rates, cost anomalies, or safety filter triggers.

Transition to Production

Plan for Integration

To take a Generative AI prototype/pilot implementation to production, it needs to integrate with mainstream workflows. It is like the building is ready, but for it to be sustainable for humans to start living, it needs water and electricity connections, security, and other social amenities. Similarly, after the Generative AI prototype/pilot is ready, it needs to be integrated with the mainstream production workflows and systems like CRM/ERP/ITSM, etc. At this time, it becomes part of the continuously running process flow and hence needs rigorous access control, continuous monitoring of system performance, and ensuring that outcomes are aligned to business KPIs.

Integration of the Generative AI prototype with other components of the business process should be done in accordance with established integration patterns to demonstrate repeatable values. Some of the most common patterns that should be followed for integration are as follows:

- **Event Driven Integration**: Event driven asynchronous integration helps orchestrate, sequence steps, and handle retries.

- **API-First Integration**: Expose Generative AI capabilities as an API that are governed, can be easily consumed, and integrated. This increases reuse across channels and simplifies integration.

- **Data Integration for RAG**: Use Vector DBs alongside traditional stores for RAG data stores. Poor data integration is one of the top reasons for pilot failure in production. Hence, implement data pipelines to ensure data quality, metadata, lineage, and incremental updates to RAG datastore.

- **Contract-Based Integration**: Integrate with downstream enterprise systems like CRM/ERP/ITSM, etc. using stable contracts to avoid tight coupling.

- **API Management and Governance**: Manage, secure API calls to LLM models, embeddings, vector search, and AI services using an API Management platform as an **AI-Control tower** with rate-limits, quota, input validation policies, authentication, and authorization. Also govern the cost by dynamic model routing, monitoring the token usage and throttling requests.

- **Separation of Concerns**: Keep Generative AI logic separate from UI or monoliths.

- **Design for Resilience**: Implement timeouts, retries, circuit-breakers, and graceful degradation to ensure that production system remains functional without AI.

- **Observability by Default**: Log prompts and responses securely with PII masked. Track token usage, latency of integration, model accuracy, and errors for enterprise visibility.

- **Cost Control**: Implement budget alerts and throttling to track model and infrastructure usage costs. This will prevent an unexpected budget overrun.

Address Change Management

Challenges in change management are among the biggest impediments to scaling Generative AI. Most Generative AI projects fail due to a lack of adoption across people and organizational processes. It fails when organizations don't change how people work, decide, and take responsibility. Below are some proven change management best practices that can help to move Generative AI pilots smoothly to production:

- **Prepare People's Mindset Early**: Educate and train people on the capabilities and limitations of Generative AI. Training on how to use Generative AI systems and setting expectations for outcomes will change people's mindsets. This will lower the resistance and drive higher adoption of Generative AI systems.

- **Define the Operative Model**: Design RACI for prompt, model, and data. Define who supports the Generative AI system and who resolves issues with models and data drift. There must be a transparent handoff from the pilot to the operations team.

- **Promote Transparent and Continuous Communication**: Sharing experience (success and failures) with Generative AI systems will help to set realistic expectations for people gradually. All should understand that the output from Generative AI is probabilistic, and not deterministic. This helps build trust, manage hype, and increase the adoption of Generative AI.

- **Reward Adoption, not Just Innovation**: Promote the use of Generative AI systems by tying them to consumers' performance goals and KPIs. It is essential to recognize those who use Generative AI solutions consistently and efficiently in their day-to-day activities, and not just those who built it. This will drive mass adoption.

- **Shift toward AI-Enabled Process**: Generative AI should not be viewed as just another Chatbot or a tool. It should be designed to embed into the business process workflows so that it is invisible yet indispensable. This will promote a seamless adoption of Generative AI solutions.

Deploy Incrementally and Scale Strategically

Adopt an incremental approach to deploying and rolling out Generative AI initiatives. This helps to gather feedback, manage risks at each stage, and reduces the chance of widespread failures. Teams can refine their approach based on the learnings before full-scale rollout. The following are some of the best practices that can be followed:

- **Plan Deployment of Generative AI Pilots Based on Their Complexity and Business Value**: Start with Tier 1 initiative that has low complexity and high business value and then plan for more complex initiatives. This helps to keep the risks of failure and setbacks low and learn incrementally before moving to more complex public-facing applications that require stricter compliance and higher governance.

- **Plan for an Incremental Rollout in Phases**: A typical progression moves from pilot with a small group of users or systems (5–20 users or instances) to a limited rollout with a larger number of users and features, and finally full production deployment.

- **Define the Functionalities, User Groups, or Geographic Locations** to be included in each phase to scale strategically.

- **Deploy the Initial Pilot to Validate Technical Feasibility and Gather Initial Feedback** from real operating conditions. Collect user feedback and incorporate learnings to improve subsequent rollouts and enhance overall system quality. Incrementally roll out the Generative AI initiatives to a broader set of external users.

- **Set the Criteria and Checkpoints to Be Met at Each Phase** before moving to the next. Monitor system performance, user adoption rates, KPIs, and business outcomes at each milestone.

- **Have a Backup and Roll-Back Strategy** in place if there are issues with the broader audience. This helps to learn and adapt to feedback and scale incrementally.

Measure ROI

Business-aligned ROI is a critical factor in determining whether a Generative AI pilot will be promoted to production. If the ROI is vague or is purely technical, the Generative AI pilot is most likely to stall. When ROI is **business-aligned, staged, and credible**, scaling decisions become easy. Hence, it is important to identify and measure the ROI of Generative AI right from the beginning of pilot implementation. Organizations should establish baseline metrics before piloting Generative AI so that any improvements can be measured objectively.

The following are some of the best practices to measure the ROI of Generative AI pilot:

- **Define the ROI of the Generative AI Use Case** from the pilot phase onward. Focus and document the expected business benefits. This will prevent interesting but unscalable pilots and secure funding and support from CXOs.

- **Measure the Cost to Serve**: Cost to serve Generative AI includes cost per query/transaction, model/token usage cost, infrastructure cost, monitoring and support cost. Include real costs for licenses, data preparation, governance and training.

- **Measure Adoption As the ROI Indicator**: Track no. of users using the Generative AI system, frequency of use, number of tasks completed, error rates to measure adoption.

- **Measure the Reduction in Risks** or the avoidance as an alternative for direct benefits to justify the ROI of moving Generative AI pilots to production. Reduced error rates, fewer compliance issues, reduced fraud losses, fines avoided by using AI to monitor transactions - are other indicators of risk reduction using AI that can justify the ROI of moving Generative AI pilots to production.

Conclusion

Scaling Generative AI pilots to production requires careful and early planning. More than selecting the right LLM models, it needs extensive considerations for technical infrastructure, organizational readiness, and risk management. Pilots should not be treated as throwaway experiments.

They should be treated as a learning ground to lay the foundations for production-grade MLOps/LLMOps and a robust framework for responsible AI governance, ensuring the safety of the Generative AI system.

Organizations must invest in cross-functional capabilities and a phased execution strategy that validates business value at each stage while managing complexity and risk. This builds organizational confidence progressively. Start small. Learn from each phase of deployment and expand strategically for production rollout. Disciplined execution at each step is a success mantra for the successful production rollout of Generative AI.

Start by selecting a high-value pilot, apply the best practices outlined in this chapter, and use those learnings as a blueprint to expand the Generative AI initiatives across the enterprise for successful production deployments.

CHAPTER 10

The Next Wave of Generative AI: Preparing Business and Technology Leadership for the Future

Introduction

Generative AI has seen unprecedented development and adoption since ChatGPT introduced it to the mainstream in November 2022. Many organizations have been using it in various shapes and forms. It is mainly used as a tool for research, study, or content creation but still awaits deeper integration into the core business workflows. For example, several customer

© Brajesh De 2026

B. De, *Generative AI for Business Innovation*, https://doi.org/10.1007/979-8-8688-2682-5_10

service teams already use Generative AI to draft first-response emails and summarize support tickets, reducing response times, while they continue with human reviews.

According to a 2025 McKinsey survey, most organizations are still experimenting or have been using Generative AI primarily in pilot phases. Only a few companies have begun scaling their Generative AI programs across the organization. Generative AI is shifting from isolated pilots to enterprise platform. However, the number of organizations that have successfully deployed and integrated Generative AI across the organization remains remarkably low.

In parallel, organizations have also begun exploring opportunities to build autonomous workflows with AI agents. They are looking at Agentic AI, systems built on foundational models, that will orchestrate end-to-end tasks across systems. AI is moving from chatbots, powered by Generative AI, that respond to individual prompts with information or content, to Agentic systems that can act on their own and work toward long-term goals.

The next few years will see several significant changes that will reshape the industries. The rise of domain-specific models, agentic workflows, multimodal systems, real-time/edge deployments, and governance-by-design are among the key future trends that are likely to have a significant business impact. CTOs and business leaders should keep a close eye on them and be prepared to integrate them into business processes for a positive impact on our everyday lives.

Generative AI technologies will continue to evolve and drive rapid market transformation in the coming years. Massive investments by large technology companies and the research labs are accelerating innovation at an unprecedented pace. This is a significant challenge for the technology leaders to stay abreast of the latest developments.

This chapter will explore some of the major emerging trends in Generative AI. It explains the business impact, tech implications, and associated risks. It also provides guidance to CTOs, enterprise architects, and business leaders on how to prepare for the next evolution of AI.

Core Technology Shifts

The Rise of Domain-Specific Foundation Models (DSLMs)

General-purpose LLMs perform well for a broad set of use cases. However, they fall short when it comes to performing industry-specific specialized tasks. Domain Specific Models (DSLMs) provide higher accuracy, lower cost, and greater compliance as they are trained or fine-tuned with specialized datasets, text, or code for a specific industry. Domain models are generally smaller in size, need less computing power, deliver better accuracy, offer greater control, and reduce risks of hallucination.

Gartner predicts that by 2028, 60% of the Generative AI models used by industries will be domain specific. They will become the dominant deployment pattern, especially in regulated industries to improve accuracy, mitigate risks, ensure compliance, and reduce cost.

However, building domain-specific models comes with its own challenges as follows:

- Needs investments in hardware (GPU and Compute) and resources (AI/ML Engineers) to train the model.

- Data acquisition and pre-processing to train the model can be daunting.

Instead of building custom domain models, it is therefore advisable to use either or a combination of the following to adapt language models for a domain or industry:

- Use off-the-shelf domain-specific models that can be trained or tuned to meet enterprise needs.

- Use prompt engineering with targeted prompts optimized for specific tasks to steer outputs for domain tasks.

- Use Retrieval Augmented Generation (RAG) to feed enterprise-specific contextual information.

- Fine-tune LLMs with synthetically generated industry data without getting into the complexities for data acquisition and pre-processing.

Agentic AI Ecosystem Powering Autonomous Workflows

The industry is transitioning from chat interfaces to **Agentic AI**—autonomous systems that can plan, act, learn, and collaborate to complete multi-step tasks independently across digital environments. Each AI agent specializes in performing specific tasks and can become a digital teammate. Frameworks like LangChain and AutoGPT are pioneering this. When implemented successfully, AI can plan and book a full vacation rather than just suggest destinations. It can triage and resolve IT service tickets, adjudicate claims, and much more. The possibilities are endless.

The industry will also see growth in **Multi-Agent Systems (MAS)**, where a collection of specialized AI agents collaborate to complete complex workflows. Each of these agents will perceive the environment, reason through a task, collaborate with each other, and handle non-deterministic outcomes without human intervention. This will improve overall accuracy, efficiency, and scalability compared to monolithic AI solutions.

Gartner predicts that 70% of MAS will use narrowly specialized agents by 2027. These agentic systems will evolve over time. They will move from a single platform to cross platform to "internet of agents" that can dynamically discover and interact with each other. Figure 10-1 illsutrates the expected evolution from isolated deployments on a single platform to interconnected multi-platform ecosystems and eventually an Internet of Agents.

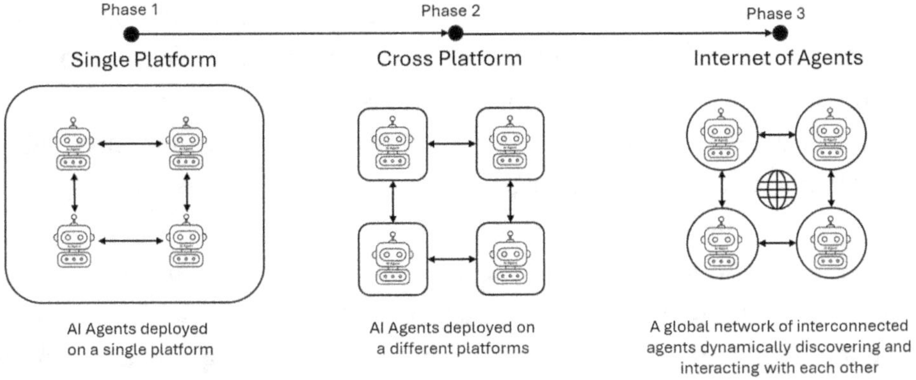

Figure 10-1. *Evolution of AI Agents*

As the Multi Agent ecosystem evolves, **CIOs will need to establish governance for agent interoperability, security, and compliance**. Leaders will manage a hybrid workforce in which humans and agents collaborate. Communication of change management plans to address workforce concerns will become important.

Multimodal and Contextual Intelligence

Just like human brains can analyze different sensory inputs to have a holistic understanding of the world and perceive reality, a multimodal Generative AI model can integrate and interpret multiple data types within a single model. A multimodal Generative AI model supports multiple data types like text, audio, images, and video, which allows them to interact with the world in innovative, transformative ways. GPT-4V, Gemini-2.0 are examples of popular multimodal Generative AI models that support text, image, audio, and video.

By combining information from different types of content, a multimodal AI can better understand it. This helps them to process more complex queries and also respond with fewer hallucinations. As a result, multimodal Generative AI can be used to implement more complex use cases in the

industry. For example, a multimodal Generative AI model can process
and integrate multisensory information from IoT devices to deliver more
personalized experiences to customers across industries. Users can
communicate with such models using gestures and voice commands. This
ease of interaction enables more people with diverse abilities to engage with
Generative AI and reap its benefits.

With the gradually increasing power, capability, and accuracy of
Generative AI models at reduced costs, the field of multimodal Generative AI
is rapidly evolving. New and innovative use cases are being built using these
models, reshaping the endless possibilities of AI. For example, by blending
text, image, and videos, an organization can create more engaging marketing
campaigns. For example, a cosmetic brand can upload product photos and
reference campaigns to generate concept boards and video advertisements
for social media in different styles. Similarly, virtual assistants can
dramatically transform patient care by communicating in different modes
like text, speech, images, videos, and gestures, making interactions more
intuitive, empathetic, and personalized.

Enterprise and Architecture Trends

AI-Native Platform Standardization

Enterprises that began with pilot implementations of Generative AI
are moving toward standardizing their AI technology and observability
stack—meaning systems that track model performance, drift, cost, and
safety telemetry. They are looking to deliver results quickly with AI-native
development platforms to boost speed and reduce costs. AI-Native software
development platform uses Generative AI to develop software faster and
more easily. Software engineers will use these AI-Native platforms to
develop applications.

Vibe-coding tools are gaining popularity as they enable software development without deep technical knowledge. However, they are good for developing prototypes till they mature to take care of aspects that are needed to build scalable applications like modularity, security, logging, monitoring, and accessibility.

While CEOs and CFOs are focused on cost savings, CTOs and CIOs are more interested in improving software development productivity through AI-native development platforms. This will shift capacity to higher value work. According to Gartner, 80% of organizations will transform large software engineering teams into smaller, AI-augmented teams by 2030. These teams will work with AI-powered tools and platforms to develop applications at a much faster pace than today while also improving quality.

Modular and Composable Architecture

Latency, sovereignty, and resilience of Generative AI applications are driving the use of microservices and APIs for building modular AI systems that can adapt quickly to changing needs and integrate seamlessly with legacy systems. A modular approach not only provides flexibility but also makes the architecture future-proof. It provides the following benefits:

- **Flexibility and Future-Proofing:** Models, vector databases, prompt engines and evaluation frameworks can be swapped easily without touching downstream orchestration or any significant rewrites. Composable design allows the use of AI services from any vendor or choice, namely, AWS Bedrock, Google Vertex, Azure OpenAI. This avoids vendor lock-in and protects the architecture against pricing-shifts, licensing changes, or availability issues.

- **Rapid Experimentation and Innovation:** Modular
  design enables rapid experimentation by easily
  swapping embeddings and evaluating prompts without
  rewriting pipelines.

- **Reusability across Use Cases:** Components like
  tokenization, embedding, retrieval, safety filtering, and
  content moderation are reusable and can be easily
  plugged into different use cases. This helps to rapidly
  build new Generative AI use cases at scale.

- **Interoperability with Enterprise Systems:** A modular
  design helps easily integrate with enterprise systems
  such as CRM, ERP, ITSM, MDM, and others. It also
  facilitates reuse of enterprise assets and enables event-
  driven architecture using Kafka, Pub/Sub, or EventHub.

- **Performance Optimization and Cost Control:** Since
  a modular design helps to optimize at each layer, it
  reduces latency, GPU cost, inference overload, and
  network bandwidth.

Critical workloads are generally implemented using an event-driven
architecture deployed at the edge for real-time communication.

Hybrid and Multi-Cloud AI Infrastructure

A hybrid and multi-cloud architecture for Generative AI is necessary
for enterprises to balance cost, data locality, scalability, and security.
Enterprises are adopting flexible hybrid cloud strategies and leveraging
specialized hardware (GPUs, TPUs) to manage the intense computational
demands of AI workloads, balancing on-premise needs with cloud
scalability. They are running sensitive workloads on private/on-prem
(hybrid) cloud and using public cloud for compute resources needed for
training, fine-tuning, and inference.

A hybrid cloud architecture provides the necessary scalability, performance, control, and access to specialized AI services needed to develop and deploy powerful GenAI models, integrating private infrastructure with multiple public clouds for a unified platform. A hybrid AI deployment integrates on-premises infrastructure with public or private cloud services to seamlessly move data and applications between these environments.

A multi-cloud deployment architecture allows organizations to distribute their AI workloads on multiple cloud vendors. This prevents vendor lock-in and optimizes performance by selecting the best service from different service providers.

Hybrid Cloud for Generative AI

The core drivers for a Hybrid Cloud for Generative AI are as follows:

- **Regulatory, Data Sovereignty, and Compliance Pressures**: Industries like BFSI, health care, and defense have strict regulatory requirements to keep data within the country, state, or organizational boundaries. Sensitive and high-value data like PII, PHI, financial transactions, and intellectual property require on-premise control. Firms need to comply with regulations like GDPR, HIPAA, PCI-DSS, FedRAMP, ISO 27001, DPDP Act, which makes full public cloud deployment difficult. In such cases, hybrid cloud deployment is a safe compromise.

- **Scalability**: Generative AI workload that needs high computational power can be executed on public cloud infrastructure to leverage the massive power and scale cost effectively.

- **Data Centricity**: Running AI workloads closer to data sources reduces latency and improves performance.

Multi-Cloud for GenAI

The core drivers for a multi-cloud deployment of Generative AI systems are as follows:

- **Best of Breed Services**: Model diversity, processing power, and storage drive multi-cloud deployments. Every model has its strengths. Enterprises want to selectively route their Generative AI requests to different clouds to take advantage of cutting edge and best of the breed AI/ML services provided by each cloud provider. This is also driving innovation.

- **Avoid Vendor Lock-in**: An abstraction layer reduces dependency on a single cloud provider. This enables easy switching between Azure OpenAI, Google Gemini, AWS Bedrock, or OnPrem SLM stack. Organizations generally use an AI abstraction layer with AI Gateway or frameworks like LangChain to easily swap out underlying models with minimal code changes.

- **Resilience and DR**: A multi-cloud deployment of AI workloads improves the disaster recovery capability and enhances resilience of the system by distributing different workloads across different cloud platforms.

Smaller, Specialized, and Efficient Models

There has been a surge in small language models (SLMs) like Microsoft's Phi-3, Google's Gemma, and Meta's Llama series. They are cheaper to run, smaller in size, optimized for speed, and can be deployed on device, often fine-tuned for specific tasks. They offer better control/accuracy for enterprises, reducing "hallucination" risks in critical domains.

They can be used in highly regulated sectors like Banking, Financial
Services and Insurance (BFSI) for building customer support Bots that can
handle common queries like balance checks, KYC status, card blocking,
claims. They can easily summarize internal policy documents or regulatory
circulars from governing bodies like RBI/IRDA, etc. They can be deployed
on premises, offering higher security and control.

In the Healthcare and Life Sciences domain, they can be used to convert
messy dictation into structured SOP notes on device with zero leakage of
patient data. It can help to streamline data entry for pre-
authorization required for Insurance purpose.

Data Governance and Security Trends

A robust data foundation and effective data governance are critical enablers
of sustainable AI. Weak data governance practices make Generative AI
models vulnerable to bias, disinformation, regulatory breaches, and security
risks. Gartner predicts that 60% of the organizations will fail to realize the full
value of AI due to inadequate data governance.

Generative AI is transforming data management and governance
practices by enabling automation and proactive risk management.
Companies that adopt these practices will gain real advantages through
faster decisions and better customer experiences.

"AI-First" Data Governance

An AI-First approach is shaping the future of data governance, where many
governance activities are executed by machines in real time. Generative AI
is increasingly used to automate routine data engineering tasks like data
integration, data cataloging, data cleansing, and detecting unusual patterns.
It can also translate legal or policy documents into enforceable technical
rules that can be applied consistently across data platforms.

Leading platforms like Snowflake and Databricks use AI to organize, classify, and monitor data continuously at scale. AI is used to detect data issues before they affect business operations. This allows organizations to act proactively rather than reactively.

As a result, data governance is becoming real-time and context-driven as policies adapt based on factors like user role, purpose, geography, current risk level, model type. Generative AI is being used as "co-governor" of data, while AI agents act as intelligent data stewards that automatically detect policy violations and recommend fixes.

The current trend shows the adoption of Generative AI for the following data governance activities:

- **Data Cleansing:** Generative AI can identify and fill gaps in an incomplete dataset with realistic synthetic data points. It can also identify incorrect data and flag it for deletion or even autocorrect.

- **Data Standardization:** Apply standardization rules based on specific requirements to align data standardization processes. It can normalize formats, convert units, and even fill missing values intelligently using contextual inference.

- **Data Quality:** Generative AI models can more efficiently understand patterns and detect anomalies, inconsistencies, and data quality issues than humans. This helps to remove biases, reduce hallucination, and improve the accuracy of AI system.

- **Data Deduplication:** Generative AI can compare data not just at the textual level but also at the semantic level. This helps to recognize and link data that represent the same entity and remove duplicates.

GenAI-Enhanced Threat Detection and Insider Risk Management

Today, all organizations are highly focused on securing their structured data. With the rise of Generative AI, Gartner predicts that organizations will reprioritize their security efforts toward unstructured data security initiatives. Generative AI will enhance threat detection and strengthen security analytics. AI models summarize and correlate logs from SIEMs, DLPs, EDRs, data platforms, and API gateways, and identify subtle patterns such as data exfiltration or misuse of privileged access. It can quickly identify deviations from the regular patterns and highlight anomalies in data.

Business and Workforce Implications

Generative AI as the Copilot and Force Multiplier

Generative AI is like scaffolding that amplifies human potential, helping people do more and become better. The real impact is not going to be the power of the AI models, but how humans choose to use AI to achieve their goals. People will have to take into account the societal and environmental impact of AI and use it accordingly. A lot will depend on how we use our energy, compute, and talent toward the development of AI models and solutions.

Generative AI agents will be re-engineered to improve scalability and efficiency of workflows. Powered by the company's proprietary data, these agents will improve context-aware decision-making. Agentic AI solutions will automate and orchestrate workflow solutions, making them efficient and trustworthy. Acting as co-pilots, these agents will improve human productivity and provide better customer experience.

However, there are risks with AI agents around transparency, ethics, and security. A pragmatic approach with AI agents working collaboratively with

human decision-makers and subject matter experts will make the agentic
AI solutions more trustworthy. The benefits and risks of every agentic AI
solution should be weighed in with a human-centric approach. There has to
be a balance between "done-for-you" experience and giving the end users
the authority and control to take the final decision. Hence, an agentic AI
solution must be built with a seamless blend of advanced autonomy and
user-centric controls that brings in transparency, ethics, and the current
context to make the applications more scalable and trustworthy.

Physical AI

Physical AI will bring digital intelligence into robots, wearables, drones,
vehicles, and smart devices. Powered by sensors, actuators, and AI models,
these physical devices can sense, decide, and act. They will become our
digital twins.

These physical devices may also be integrated and interact with legacy
systems. CIOs will have to ensure safety, reliability, and explainability of the
decisions and actions taken by these devices. This will need adherence to
the highest standards of safety and regulatory compliance.

The Green Price of Intelligence: Environmental Risks of Generative AI

AI Technology companies have been building and training Generative AI
models on huge volumes of data. This has led to a surge of infrastructure
demand that the world has never seen before.

- As per the International Energy Agency (IEA), traditional
  enterprise data centers required **10–20 MW.** But Modern
  AI-ready sites require **100–300 MW.**

- Hyperscale campuses were approaching **1 gigawatt**—power equivalent to feed around **800,000 homes**.

- Data center power requirements in North America **jumped from 2,688 MW at the end of 2022 to 5,341 MW at the end of 2023**, driven heavily by Generative AI workloads.

- By 2026, global data center electricity consumption is expected to reach **1,050 terawatt-hours (TWh)**—more than the total energy consumption of some countries.

Training GPT models is not just expensive but also resource intensive.

- As per MIT Technology review, GPT-4 model consumed ~**50 gigawatt-hours** only for its training—enough to power **San Francisco for three days**.

- Researchers at UC Berkeley estimated similar model training used **1,287 MWh** and emitted ~**552 tons of CO_2**.

- A single GenAI prompt consumes 10 to 100 times more energy than a simple search.

This is not the end. Continuous fine-tuning of models continues to draw energy. Retraining new versions of larger, data-hungrier, parameter-heavy Generative AI models consumes more power than earlier. Continuous use of Generative AI models increased GPU usage, thereby increasing the carbon footprint.

Generative AI is not just a software revolution—it's an **energy revolution**. Generative AI might improve productivity, accelerate innovation, and save millions of lives. But at its current trajectory, it may also make the planet less livable. Organizations should therefore evaluate AI initiatives not only on business value, but also on energy efficiency, carbon footprint and sustainability impact. The companies that embrace both intelligence and sustainability will earn the trust—and the future.

Is Innovation Without Sustainability a True Victory of Technology?

True leaders must introspect and ask

- What is the true total cost of ownership of Generative AI initiatives—including infrastructure, data, talent, governance, and risk?

- Is our AI roadmap aligned with our sustainability commitments and ESG targets?

- What is the environmental impact of our AI usage, and how does it affect our long-term operating model and brand responsibility?

Generative AI's Future: Responsible, Explainable, Governed

Generative AI, LLMs, and foundation models will boost productivity, augment human capabilities, and unlock new growth opportunities. Companies will find new and efficient ways to do work. Every role in all organizations will be reinvented with AI co-pilots assisting humans at every step. Companies will have to make significant investments to retrain their workforce to work efficiently with AI technologies. AI co-pilots will become the norm, significantly enhancing human accomplishments. Every job will undergo a transformation, where Generative AI will either automate part of the job or assist humans in accomplishing it faster. It will also create new jobs for humans to take care of, especially to ensure responsible and ethical use of AI-powered systems.

The future of Generative AI lies in building a responsible, explainable, and governed framework that will not only transform every human and machine interaction but also ensure the safety and well-being of mankind. Every AI system must be built on four non-negotiable pillars: **policy, controls, monitoring, and audit**.

- **Policy** defines the ethical principles like fairness, privacy, security, and accountability, ensuring AI benefits humanity while minimizing harm.

- **Controls** ensure that models operate within approved guardrails.

- **Monitoring** tracks performance, costs, drift, risks, and real-world impact.

- **Audit** provides transparency into how decisions are made and ensures compliance.

Every "black box" decision by the AI systems must be explainable. AI systems must be supported by tools and methods to understand how and why a decision was made. There must be strict policies, standards, and continuous monitoring of AI systems. Only then can organizations ensure that Generative AI aligns with human values and legal requirements—building trust while delivering both societal and business value. AI must be democratized for all with complete safety, security, and ease of use.

Future Readiness Checklist for CTOs and CIOS

Area	Key Questions
Agentic AI	Are we prepared to govern autonomous agents?
Models	Do we have a strategy to adopt domain-specific models?
Architecture	Is our architecture scalable and flexible to support multimodal and agentic systems?
Governance	Is AI Governance embedded by design?
Workforce	Are our teams ready for AI-Augmented work?
Sustainability	Are we measuring AI's environmental impact?

Conclusion: Strategic Imperatives for Leadership

Generative AI is probably one of the most impactful technologies of the century. It is shifting from experimentation to enterprise transformation. As we look ahead, the future of Generative AI holds great promise. It will definitely boost productivity, fuel innovation, drive creativity, and offer personalized experiences. There will be a fundamental shift in how businesses create value and compete. Successful organizations will be those that move beyond isolated pilots and treat GenAI as a core innovation engine.

However, the power of Generative AI comes with significant responsibilities to protect society and humanity from its possible misuse. If left unchecked, it risks spreading misinformation through deepfakes, committing sophisticated financial fraud by compromising cybersecurity, amplifying societal biases, and compromising privacy. Constant use of Generative AI can also cause systemic dependency in the long run.

Organizations must ensure thoughtful governance, transparent design, and meaningful human oversight to harness the immense benefits of Generative AI, while keeping it aligned with human values. CEOs and technology executives' success with Generative AI depends on excelling in three areas:

- **Strategic Foresight:** Reinvent workflows with Generative AI to add real value.

- **Platform Thinking:** Build a modular technology, strong governance, and the right talent to scale.

- **Responsible Leadership:** Guide change ethically and transparently for long-term sustainable impact.

Leaders who treat Generative AI as a strategic capability and not as a tool will shape new markets. Companies that align culture with technology will scale AI faster.

The mandate is clear: Stop experimenting and start scaling—**Prioritize** Generative AI initiatives, **Industrialize** with a secure and governed AI platform, and **Operationalize** to full production with clear KPIs and executive accountability.

The winners will be those who combine strategic foresight, platform thinking, and responsible leadership. Markets will be led by those who act now—not those who wait to be ready.

Index

© Brajesh De 2026
B. De, *Generative AI for Business Innovation*, https://doi.org/10.1007/979-8-8688-2682-5

H

Leadership, 144

Legal professionals, 44

Life sciences, 81

LLM Ops pipelines

    automated evaluation, 297

    governance, 295

    production-grade CI/CD, 296

    prompts, templates, chains, 296

    RAG data, 296

    security & compliance, 297–299

Loan and product

    recommendation, 85

Loyalty programs, 105

M

Machine learning (ML), 3, 4,
    260, 265

Managed services, 292

Market trading, 127

Marketing, 32–33

Marketing campaign, 78

Master data management
    (MDM), 167

Media and entertainment

    game design, 97

    image and video, 97

    movie editing, 98

    music creation, 97

    music recommendations, 99

    personalized content
        recommendation, 99

    personalized gaming
        recommendation, 99

    personalized targeted
        advertising, 100

    preproduction, 97

    retail and eCommerce, 100–108

    script writing, 96

Medical imaging, 79

Metadata, 290

Microservices, 287

Minimal Viable Product (MVP), 294

Misrepresentation, 55

Mitigation, 72

Mitigation strategies, 262

MLOps, 226–229, 261

Movie production, 98

Multi-agent systems (MAS), 314, 315

Multi-cloud deployment
    architecture, 319

Multi-cloud for GenAI, 320

Multimodal Generative AI
    model, 315

Multisensory information, 316

N

Natural language processing (NLP),
    14, 23, 88

Neural networks, 22

Non-negotiable pillars, 326–327

Non-production time (NPT), 124

O

Observability, 305

Off-the-shelf Generative AI, 278

Q

R

S

T

GPSR Compliance
The European Union's (EU) General Product Safety Regulation (GPSR) is a set
of rules that requires consumer products to be safe and our obligations to
ensure this.

If you have any concerns about our products, you can contact us on

ProductSafety@springernature.com

In case Publisher is established outside the EU, the EU authorized
representative is:

Springer Nature Customer Service Center GmbH
Europaplatz 3
69115 Heidelberg, Germany